# The Healthy
# Thyroid

# The Healthy
# Thyroid

What you can do to prevent
and alleviate thyroid imbalance

Patsy Westcott

Thorsons
An Imprint of HarperCollins*Publishers*
77–85 Fulham Palace Road,
Hammersmith, London W6 8JB

The website address is: www.thorsonselement.com

and *Thorsons* are trademarks of
HarperCollins*Publishers* Ltd

First published as *Thyroid Problems* by Thorsons 1995
This revised edition published by Thorsons 2003

1 3 5 7 9 10 8 6 4 2

© Patsy Westcott 2003

Patsy Westcott asserts the moral right to
be identified as the author of this work

A catalogue record of this book
is available from the British Library

ISBN 0 00 714661 2

Printed and bound in Great Britain by
Creative Print and Design (Wales), Ebbw Vale

# Contents

# Acknowledgements

Many people have helped with this book. I should like to thank Dr Mark Vanderpump, secretary of the British Thyroid Association, for his help with queries, and Lyn Mynott of Thyroid UK and Lyn Welch of the Thyroid Eye Disease Association, for sharing their ideas and experiences. Many thanks also to the British Thyroid Foundation and the Thyroid Eye Disease Association for supplying leaflets and information.

Above all, a large and heartfelt thank you goes to all of the women who so honestly and willingly shared their personal experiences of thyroid disease – without your contributions, this book would not have been possible.

# CHAPTER ONE

# The Hidden Illness

Thyroid disease is common and affects women more
frequently than men.

Many books and articles on thyroid problems for both the
general public and medical profession begin with these or similar words. But this bland statement barely begins to suggest
the number of women afflicted by thyroid problems or the
impact of thyroid disorders on our lives. In fact, according to
a review in the *British Medical Journal*, taken together, underactive and overactive thyroid conditions represent the most
common hormonal problem – and this problem overwhelm-
ingly affects women.

In terms of statistics alone, thyroid problems in women
deserve to be taken seriously:

- Four out of five people with thyroid disorders are women.
- One in 10 women will develop a thyroid disorder at some
  stage in her life.
- Between one and two in 100 women in the UK will
  develop an underactive thyroid (hypothyroidism), a
  condition 10 times more common in women than men.
- Two out of every 25 women – and one in 10 past the
  menopause – have so-called mild thyroid failure that is
  considered borderline on blood tests. These 'subclinical'
  problems are linked with nagging ill health, such as
  fatigue, mood swings and overweight, as well as more

serious medical problems such as depression, heart disease and osteoporosis.

- Overactive thyroid conditions (hyperthyroidism) are also common in women, affecting between five in every 1000 to one in 50 – or 10 times more women than men.
- One in every 100 people in the UK will develop an autoimmune thyroid disorder, when the body turns against itself to cause the thyroid to become either underactive or overactive. Autoimmune disorders, including those affecting the thyroid, are estimated to be the third biggest killer after heart disease and cancer.
- Having a personal or family history of autoimmune disorders, such as diabetes or rheumatoid arthritis, gives you a 25 per cent greater risk of developing a thyroid disease than someone without such a history.
- Hashimoto's thyroiditis – an autoimmune disorder causing an underactive thyroid – may account for up to one-third of such cases in this country and is five times more common in women than men.
- Graves' disease – an autoimmune condition causing an overactive thyroid – is 15 times more likely to affect you if you are a woman.
- Goitre (a swollen or enlarged thyroid gland) is four times more common in women than in men.
- Thyroid nodules or lumps are also more common in women – estimated to affect about one in 20 women.
- Thyroid cancer, although rare, is also more likely to develop if you are a woman.

## Only as Healthy as Your Thyroid?

Thyroid problems can affect a woman at any age or stage in life – from the teens to retirement. Throughout this time, they are a source of much ill health and unhappiness. During the reproductive years and after the menopause, they can exacerbate other female health problems as well as create a host of debilitating symptoms that affect every system of the body:

- Thyroid problems can cause menstrual disturbances, such as heavy or absent periods, and worsen problems such as premenstrual syndrome (PMS).
- Thyroid problems are an underrecognized cause of fertility problems and miscarriage.
- During pregnancy, thyroid disorders are the most common hormonal problem.
- Even a mild shortage of thyroid hormone during pregnancy may affect the unborn child's future IQ (intelligence quotient). Research shows that children, aged seven to nine, whose mothers had untreated hypothyroidism during pregnancy scored about seven points lower on IQ tests.
- According to US research, women with faulty thyroid function are more likely to give birth to babies with defects of the heart, brain or kidney, or have abnormalities such as a cleft lip or palate, or extra fingers.
- Babies whose mothers have an underactive thyroid have an increased risk of heart problems – even if their mothers are being treated for the condition. Yet, at the time of writing, the NHS still does not routinely test thyroid function in pregnancy.
- One in 10 young women have thyroid problems after giving birth, with symptoms such as depression, tiredness and a lack of zest that cast a shadow over the first months of parenthood. Such symptoms are often misdiagnosed as 'the baby blues', thus depriving women of treatment that would help.
- Later in life, thyroid disease becomes even more common. An estimated one in 10 women over 40 may have undiagnosed thyroid disease, which is particularly worrying as thyroid problems are associated with an increased risk of two very significant causes of female ill health in later life: heart disease and osteoporosis (brittle-bone disease).
- One in five women over 60 suffer thyroid problems. With the 'baby-boomers' reaching this age, thyroid disorders will become an increasingly major health challenge.

- Thyroid disease in older women is more likely to be 'silent', producing few or vague symptoms. But compared with, say, high blood pressure – another 'silent' disease with serious consequences – thyroid problems are far less likely to be suspected or tested for.

These facts and figures alone put thyroid disease on a par with conditions like diabetes, estimated to affect one to two in every 100 people, and breast cancer, which strikes one in eight women. However, thyroid problems attract only a fraction of the research funding given to these high-profile conditions, and have until only recently relatively poor media coverage. That this is now beginning to change was reflected by an editorial in the prestigious British medical journal *The Lancet* that declared 'you're only as healthy as your thyroid'.

## Are Times Changing?

During the writing of the first edition of this book more than seven years ago, there was little awareness – even among journalists specializing in women's health – of just how common thyroid problems are and of the misery they can cause. A small request for help placed in *The Guardian* newspaper resulted in a deluge of phone calls: 200 over two days.

Revisiting thyroid problems now, has anything changed? The good news is that there has been a shift in knowledge and attitudes. A great deal more is becoming understood in terms of how thyroid problems are caused and how they may best be treated. And certainly, many more people are now aware of thyroid disease than in 1995.

Part of this new awareness is thanks to a number of books drawing attention to the wide-ranging effects of thyroid problems and the misery they can cause. The advent of the Internet has also done much to fill the information gap. There are now several excellent websites where women with thyroid problems can get information and communicate with others who have the same condition. This is good news for the millions of women living with a faulty thyroid.

However, in other aspects, the changes have been pitifully few. Thyroid disease is still a 'Cinderella' disorder, despite being the cause of so much depression, tiredness, discomfort and feeling well below par. Even nowadays, women often soldier on for long periods before anyone takes their complaints seriously – hardly surprising given that the average medical student only has a lecture or two on thyroid problems, if they're lucky. And although there are more post-training courses for interested doctors, endocrinology (the study of hormones) is still not a core subject in most basic medical courses.

Like their 'sisters' in 1995, many of the women interviewed for this new edition had struggled on for months, even years, with crippling symptoms before being diagnosed. Once diagnosed, they had to cope with unsympathetic doctors and endure treatments that were uncertain, took time to get right and sometimes didn't work at all.

Just like seven years ago, many women also spoke of the dilemmas posed by treatment – how long it had taken for medications to start working, the uncertainty of their effects, the agony of deciding whether to opt for surgery or radioiodine therapy, or whether different forms of medication might be more effective. All related stories of how difficult it was to obtain relevant information and how alone they felt with this unpredictable disease.

Others told poignant stories of how the disease had affected their daily lives and relationships in the face of the pressures of holding down a job while battling overwhelming fatigue, the difficulties encountered with partners, friends and children who did not always understand why the person they loved had undergone such a major personality change, of being overweight or underweight, of the self-consciousness endured because of bulging eyes, thinning hair and thickened skin, and the effects these had on their self-esteem. Some described the heartache of not being recognized by friends they had not seen for some time.

These physical problems are often dismissed as trivial but, in a world where the pressure to look young and attractive is intense, they can become the cause of a huge amount of distress and self-loathing. They can also lead to other women's health problems such as eating disorders.

## Tip of the Iceberg?

In recent years, it has become apparent that thyroid problems may be even more common than ever imagined. The availability of more sophisticated methods of testing thyroid function has revealed that many seemingly healthy women with apparently normal thyroid function have, in fact, abnormal levels of hormones and antibodies against the thyroid gland. This also suggests that the cases of thyroid disease identified and treated may be only a fraction of what is actually out there.

Since 1995, such cases of mild or low-grade thyroid disease have received a great deal of attention in the medical and public media. Medical journals, such as the influential *New England Journal of Medicine*, have carried major articles on subclinical thyroid disease, while the shelves of bookshops now carry stacks of books about thyroid problems aimed at the general public. Many put forward the view that hidden thyroid problems are a factor in a host of conditions reaching epidemic proportions in the 21st century, including:

- Chronic fatigue
- Depression, anxiety and mood swings
- Difficulty in losing weight
- Eating disorders
- Menstrual problems
- Fertility problems, miscarriage and premature births
- Perimenopausal and menopausal problems
- Changes in libido
- Heart disease
- Osteoporosis
- Ageing.

There is much controversy surrounding the issue of mild thyroid disease – or should it be called early or pre-thyroid disease? Despite being more widely recognized, there is little consensus on its significance and whether or how it should be treated. No one truly knows how important it is as a cause of ill health or how often it might lead to full-blown thyroid disease. Other questions remain, too: Should it be tested for in the absence of symptoms? Should women with symptoms suggestive of thyroid problems, such as tiredness and depression, be treated even if blood tests are apparently normal? Should widespread thyroid screening be introduced and, if so, at what age and how often should testing be performed?

Just as in 1995, there is a lot of debate, but no definitive answers.

## Why Me?

Almost every woman included in this book wanted to know why she, in particular, had developed a thyroid disease. Unfortunately, there are no simple answers to this question. Despite its prevalence, the experts themselves still do not fully understand what causes the thyroid to misbehave. As with so many illnesses, one of the most pressing questions is whether nature or nurture lies at the root of the problem.

The cracking of the human genome, the inherited 'database' of some 40,000 to 50,000 genes containing all of the instructions for life, has led to an explosion of genetic research – and some interesting insights into the origin of certain kinds of thyroid problems. Some of the genes involved in certain kinds of thyroid cancer have been identified as well as other possible 'candidate' genes that may lead to an increased risk of developing an autoimmune thyroid disease.

However, although genes undoubtedly play a role, they are not the whole story. As with any illness with a genetic component, possessing one or more of these predisposing genes may give you a tendency to develop a particular problem – in this case, thyroid disease – but it is your environment and individual lifestyle that may yet determine whether you actually will.

## The Immune Connection

The role played by the immune system in triggering a number of thyroid disorders remains a controversial topic. Autoimmune thyroid disease, which underlies both hypo- and hyperthyroidism, is caused by failure of a fundamental mechanism: the body's ability to recognize its own organs and tissues as belonging to itself.

If the body fails to recognize itself, it produces self-attacking proteins – known as autoantibodies – to destroy its own tissue. Experts are becoming increasingly aware of a number of diverse thyroid problems due to the production of such autoantibodies, including:

- Hashimoto's thyroiditis, which causes an underactive thyroid
- Graves' disease, which causes an overactive thyroid
- Myxoedema, or generalized swelling of the skin and other tissue
- Subclinical hypo-/hyperthyroidism, mild or hidden thyroid under-/overactivity
- Thyroiditis, or inflammation of the thyroid
- Postpartum thyroiditis, or inflammation of the thyroid after childbirth
- Thyroid eye disease (TED).

## Putting the Clues Together

Women, as we already know, are much more likely than men to develop thyroid disease. Many of those interviewed for this book added, almost as an afterthought, 'My mother (or sister, or daughter) has thyroid problems, too.' In particular, Graves' disease and Hashimoto's thyroiditis seem to cluster in families. In the past, this was dismissed as a coincidence. Recently, however, the new science of molecular genetics has led a number of researchers to look for an underlying inheritance factor in the development of autoimmunity. It is now generally agreed that as much as 10–15 per cent of us inherit an immune system with the potential to turn against itself.

Nevertheless, the development of thyroid problems is not just a matter of inheriting a faulty set of genes. Many people possess autoantibodies and do not go on to develop full-blown thyroid disease. In fact, it is estimated that only about one in 10 of those with an inherited tendency to develop thyroid antibodies will actually have thyroid problems.

One of the main aims of research, therefore, is to discover the possible triggers of thyroid problems. We know that the immune system can be damaged by many aspects of the 21st-century lifestyle that seem to have potential roles in the development of thyroid disease. Pollution, ageing, diet, stress, viral and bacterial infections, and habits like smoking and drinking are just some of the factors being explored by scientists in the hopes of finding out why the thyroid becomes faulty. Since 1995, much more information has been accrued on the roles these factors may play in triggering thyroid disease – and one important risk factor could simply lie in being female.

## The Hormone Connection

The thyroid gland is involved in virtually every bodily process, including those of the reproductive system, and thyroid disease is linked to a number of specifically female problems (*see Table 3.1*).

## Table 3.1
## Links between thyroid disorders
## and the reproductive system

| Hypothyroidism (underactive thyroid) | Hyperthyroidism (overactive thyroid) |
|---|---|
| *Menstruation*<br>• Premenstrual syndrome (PMS)<br>• Heavy periods<br>• Loss of periods | *Menstruation*<br>• Irregular periods<br>• Scanty flow<br>• Loss of periods (in severe cases) |
| *Fertility*<br>• Ovulation failure<br>• Polycystic ovarian syndrome (PCOS)<br>• Recurrent miscarriage | *Fertility*<br>• No major effects in mild-to-moderate cases<br>• Recurrent miscarriage in severe cases |

The first clue that something may be wrong with the thyroid gland is often when a woman consults the doctor on a 'woman's problem', such as menstrual irregularities, difficulty in getting pregnant, miscarriage, postnatal depression or menopausal symptoms. It is also increasingly recognized that thyroid problems may be confused with or aggravate the symptoms of women's problems such as PMS and the menopause.

With so many women's problems being linked to thyroid disease and, conversely, so many thyroid problems being associated with the reproductive cycle, could it be that the female hormones play a part in susceptibility to thyroid disease? The answer is most likely yes. Research suggests that the two main female sex hormones, oestrogen and progesterone, moderate the activity of the immune system – hence the preponderance of thyroid disease in women.

The involvement of hormones and the immune system could also explain why the thyroid may misbehave for the first time during pregnancy and after birth. It also provides a reason for why so many women develop debilitating postpartum thyroiditis (PPT), which is often confused with postnatal depression.

One of the most striking developments since the first edition of this book has been the increasing awareness that the brain and nervous system, the immune system and endocrine (hormonal) system, all previously thought to be completely separate systems, do not work in isolation. This has led to the development of new fields of study such as psychoneuroendocrinology and psychoneuroimmunology, which are dedicated to exploring the body–mind connection and the way in which each 'talks' to the other.

## With Women in Mind

This book is an exploration of these and other issues. Chapter 2 looks in detail at the thyroid gland and how it works to enable readers to understand the links between the thyroid and other body systems, and why – when it goes wrong – there may be such wide-ranging effects. The chapter also outlines some of the latest thinking on the immune system and the part it plays in thyroid problems.

Chapter 3 examines all the things that can go wrong with your thyroid, and explores some of the latest theories for how thyroid problems arise in an attempt to answer that nagging question, 'Why me?' There is also more detailed information on thyroid nodules (lumps) and thyroid cancer. Despite being one of the simplest forms of cancer to treat, survival rates in the UK have, until now, lagged woefully behind those of other countries.

Chapter 4 tackles the problem of getting a proper diagnosis. It includes a description of the various tests that may be performed, and explores the issue of what is normal and the difficulties involved in interpreting thyroid function tests.

Chapter 5 describes the available treatments, including medications, surgery and radiotherapy, and explains how they work, including their pros and cons. It also covers the debate over newer – and the revival of older – forms of treatment, and how you can work with your doctor to find the treatment that is right for you.

Chapter 6 covers the different ways you can help yourself, such as by paying attention to what you eat, making sure you get the right amount of exercise and managing stress, as well as how to come to terms psychologically with having thyroid disease.

Chapter 7 looks at how complementary therapies can help you manage your thyroid problems. These therapies are much more widely accepted now than when the first edition of this book was written, and many doctors and healthcare practitioners now acknowledge the part these therapies can play alongside conventional medical treatment.

Chapter 8 is devoted to thyroid eye disease, a particularly devastating condition about which too little is known, even now, and includes the still controversial issue of how it should be treated.

Chapter 9 describes how thyroid problems can affect you at different points in the female reproductive cycle, and includes important new information on how thyroid problems can affect menstruation, fertility, pregnancy and life after childbirth.

Chapter 10 looks at the problems that may be caused by thyroid disease at around the menopause and as we get older.

Chapter 11 investigates some of the major issues in thyroid disease and the advances made in our current understanding of the disorder, as well as takes a peek into the future at possible new treatments.

Finally, there is a glossary of terms relevant to thyroid disorders, and a list of books, websites and organizations that may prove helpful.

The more you know about the way your body works, the better able you will be to help yourself if something goes wrong. The objective of this book is to provide the information you need to help yourself, to work with your doctor to get the best treatment for your problems, and to feel more in control of your body and your life – something that women with thyroid problems often feel they have lost.

This book does not intend to tell you what to do or replace medical advice. There is a great deal of controversy surrounding thyroid problems – how they come about and how they should be dealt with. The main areas of debate have been outlined in this volume to give you an idea of what different experts think so that you can make up your own mind about how to live with your thyroid problems.

# CHAPTER TWO

# Understanding Thyroid Problems

To better understand what can go wrong with your thyroid, it is necessary to know something about how the gland works. This chapter attempts to reveal why we have a thyroid gland, and looks at the way the thyroid interacts with other systems of the body, including the immune system.

The thyroid is one of 10 glands that make up the endocrine (hormonal) system. From the moment we are conceived until the time of our death, our bodies are under the influence of a cocktail of hormones produced by this system. As this system is so finely tuned, when anything happens to disturb its delicate balance, the repercussions ricochet throughout the rest of the body.

The hormones produced by glands are chemical messengers that are carried around the bloodstream to act on cells and tissues that are often far from their site of origin. Their job is to ensure that we have the correct concentrations of metabolites – vital nutrients (such as sugars and fats), vitamins and minerals (such as calcium, sodium, potassium and iodine), enzymes and other factors essential to life – in the bloodstream.

Each gland has a specific function, but also works with the other glands to keep our body in a state of chemical balance (homoeostasis). One recent, exciting discovery is that not only do hormones interact with each other, but they also exchange messages with other chemicals produced by the brain and nervous system. Research is beginning to uncover more and more links between these major interacting systems, and to throw more light on the way hormones and chemicals produced by the nervous and immune systems work together. This, in turn, is helping to clarify the connection between mind and body as reflected by a diverse number of conditions, including thyroid disease.

## Key Sites

The glands themselves are situated at key locations throughout the body (*see Figure 2.1*). Together they produce over 50 different hormones – so-called 'mighty molecules' – that have widespread effects on us from cradle to grave. As hormones cannot be stored in large quantities in the glands, the brain programmes their manufacture by means of a complex biochemical cycle that uses a series of checks and balances to ensure that hormone levels are maintained according to your body's needs.

In addition to the endocrine glands themselves, other organs contain pockets of glandular tissue that produce hormones. One of these is the hypothalamus, a region of the brain that is both part of the nervous system and a gland.

Not surprisingly, with such a complicated system, things can go wrong. Broadly speaking, when a gland ceases to function as it should, it results in two categories of problems: the gland becomes underactive and produces too few hormones; or it becomes overactive and produces too many.

## Balancing the Body

The whole endocrine system is controlled by a series of 'feedback loops', which slow or stop a gland from working when enough hormone has been produced, and turn it back on again when more is needed – like a central-heating thermostat (*see Figure 2.2*). If the blood levels of any of the essential chemicals are low, special sensory cells are able to pick up a signal that

The thyroid gland produces the thyroid hormones thyroxine and triiodothyronine ($T_4$ and $T_3$), which control the heart and metabolic rate. It also produces calcitonin, which regulates the concentration of calcium in the blood.

The pituitary gland – just the size of a peanut – controls the actions of all the other glands in the body. Its anterior lobe produces growth hormone, which is involved in growth and ageing, and prolactin, which is responsible for milk production. Its posterior lobe produces antidiuretic hormone (ADH), which controls the amount of urine you make, and oxytocin, which is involved in contractions of the uterus during pregnancy, childbirth and breastfeeding. The pituitary also produces thyroid stimulating hormone (TSH), which triggers the flow of thyroxine ($T_4$) and triiodothyronine ($T_3$) from the thyroid.

The parathyroid glands produce parathyroid hormone (PTH), which regulates the turnover of calcium and phosphorus in the bones.

The adrenal glands consist of two organs. The first produces the stress hormones adrenaline and noradrenaline, which regulate heart rate and blood pressure. The second produce steroid hormones, such as hydrocortisone, which help convert carbohydrate into energy, and the sex hormones oestrogen, progesteron and testosterone (in men). These control sexual development, fertility and reproduction.

The pancreas produces insulin, which helps maintain blood sugar levels. A shortage of insulin leads to diabetes. Another hormone, glucagon, produced by the pancreas, stimulates the liver to produce glucose.

The gonads. In women, the ovaries produce the sex hormones oestrogen and progesterone, which control the menstrual cycle and reproduction. (In men, the equivalent is the prostate gland, which produces the male sex hormone, testosterone).

## Fig 2.1
## The endocrine system

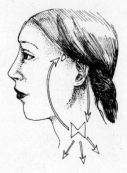

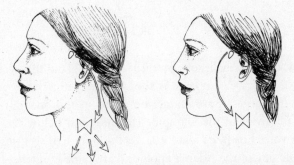

Figure 2.2 The pituitary gland and the hypothalamus in the brain work together to produce a hormone that stimulates the thyroid. The thyroid gland draws iodine from the blood in order to make T3 and T4. Sensors in the TSH-secreting cells of the pituitary detect rising levels of thyroid hormones and quell further secretion. When levels fall, the pituitary releases more TSH, which stimulates the thyroid to start making more hormones.

prompts them to release hormone. This hormone, in turn, acts on other cells to release more of the needed chemical into the bloodstream. When enough chemical has been produced, the sensory cells switch the system off, which stops further hormone release. In this way, the body's chemical balance is constantly maintained.

The system is exquisitely sensitive: food, exercise, stress, illness, changes in body chemistry such as a shortage or excess of certain nutrients, pregnancy, ageing, even the time of day or year, can affect the balancing mechanism and, with it, the amount of hormones our glands secrete.

Most hormones act only on specific tissues and not all the cells in the body. They do this by latching on to structures called 'receptors', which lie studded about the surface of or within cells, rather like a key fits into a lock. This enables hormones to be transported around the bloodstream to specific locations. Receptors are also important because, as we shall see, if the wrong chemical – such as an autoimmune antibody – attaches itself to a receptor, like a thief using a master key to get into your house, it can cause havoc and destruction.

## The Thyroid Gland

The thyroid is a small, soft, butterfly-shaped gland that weighs just 15–20 g ($^1/2$–$^3/4$ oz) and is about the size of a plum, yet it is also the largest pure endocrine gland in the body. It lies across the front of the windpipe (trachea) just below the larynx, or voice box (*see Figure 2.3*). Its two lobes, or sections, lie on either side of the Adam's apple and are joined together by a narrow bridge of tissue called the isthmus.

You may just be able to detect its outline if you look in the mirror and stretch your neck. If you take a sip of water and swallow, you may be able to see it moving up and down. If you can't see it, you may be able to feel it with your fingers. (But don't worry if you can't see or feel it – not everyone can.)

The thyroid develops in the womb during the first weeks of life from a small piece of tissue at the root of the tongue. As the fetus grows, the tissue moves down the neck to rest at its adult position. By the time the fetus is just 12 weeks old, the thyroid has already started to work.

The thyroid is made up of two types of hormone-secreting tissue: follicular cells and parafollicular cells. The follicular cells, which make up the greater part of the thyroid, are hollow spheres surrounded by tiny capillary blood vessels, lymphatic vessels and soft connective tissue. Each follicle is filled with a yellow, semifluid, protein-containing material called thyroglobulin (TG) which, when broken down, interacts with

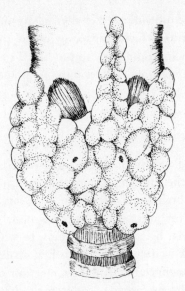

**Figure 2.3**
**The thyroid gland lies across the windpipe in the throat.**

iodine stored in the thyroid to produce thyroid hormone (TH). The parafollicular cells lie on their own or in small clusters in the spaces between the follicles and secrete another hormone – calcitonin.

## A Tale of Two Hormones

The thyroid's main purpose is the production, storage and release of thyroid hormone. Although referred as a single entity, there are, in fact, two thyroid hormones: thyroxine ($T_4$) and triiodothyronine ($T_3$), which carry four and three atoms of iodine, respectively.

$T_3$ is four times more potent than $T_4$ and works eight times more quickly. Yet, $T_4$ is about 50 times more abundant in the bloodstream than $T_3$. This is because, although small amounts of $T_4$ are converted to $T_3$ within the thyroid gland itself, most

$T_3$ is produced outside of the thyroid by a process called monodeiodination, which strips away one of the iodine atoms from $T_4$. This allows the body to produce $T_3$ as needed – like changing your five-pound notes into one-pound coins for the parking meter.

## All Under Control

Like all of the glands in the body, the thyroid is regulated and controlled by the pituitary gland, the small, pea-sized gland attached to the brain often referred to as the 'master gland'. The pituitary gland orchestrates the entire hormonal symphony but is, in turn, driven by the hypothalamus, to which it is joined by a short stalk of nerve fibres. This hypothalamic–pituitary–thyroid gland connection is a key junction through which chemical messages are carried to and from the brain and the body.

Levels of thyroid hormone are regulated by a feedback loop that operates between the hypothalamus, pituitary and thyroid glands. Anything that increases the body's need for energy – such as a fall in temperature or a bout of exercise – will provoke the hypothalamus to secrete a chemical messenger called thyrotropin-releasing hormone (TRH) to trigger the pituitary to secrete a messenger chemical called thyrotropin or thyroid-stimulating hormone (TSH), which stimulates the thyroid to secrete thyroid hormone. As the thyroid releases increasing amounts of TH, chemical messages are eventually passed on to the hypothalamus to inhibit production of TRH and, in turn, TSH.

This chemical round-robin means that TSH levels are a highly sensitive indicator of thyroid activity and can provide an early clue that the thyroid is not working as it should. This is why the TSH test is a key investigation in checking the health of your thyroid (*see Chapter 4*).

## The Calcium Connection

The thyroid is crucial in maintaining the strength and density of our bones. The parafollicular cells of the thyroid produce the hormone calcitonin, which is involved in regulating calcium levels in the body. As well as being the main mineral used for making bone, calcium is needed to trigger impulses in nerve and muscle cells.

Calcitonin acts with another hormone – parathyroid hormone (PTH) – produced by the parathyroid glands, four tiny glands that lie behind the thyroid. Whenever calcium is needed, PTH raises the levels of calcium in the blood by stimulating the release of calcium from bone, increasing the reabsorption of calcium from the kidneys and converting vitamin D into a hormone that increases gut absorption of calcium. Once calcium levels have been increased, the thyroid releases calcitonin to suppress the release of calcium from bone.

## The Incredible Thyroid

Although the thyroid is only a small gland, the hormone it produces is responsible for an incredible number of biological processes. In fact, it would be fair to say that TH is essential for the health of virtually every cell in your body. Cell growth, muscle strength, body temperature, appetite, cholesterol levels, mood and memory all depend on thyroid hormone. Likewise, your heart, liver, kidneys, reproductive organs, hair and skin all require TH to function properly.

So widespread is the activity of TH that, indeed, in Victorian times, doctors believed the thyroid was vital to life. Although it is possible to live without a thyroid, provided you receive thyroid-hormone replacement therapy, they were not that far from the truth. Thyroid hormone is unique in that, throughout the whole of our lives, it acts within almost every tissue in the body and is vital for general health and wellbeing.

## A MATTER OF ENERGY

The main job of thyroid hormone is to regulate your metabolism – the rate of your body's cell activity. It does this by activating mitochondria, the tiny cellular 'powerhouses' that produce energy. The process of metabolism – the word literally means 'change' – among other things, controls your appetite and maintains your body temperature, whatever the external environment.

Your metabolism determines the rate at which your cells burn oxygen, a process involved in every activity of life – from breathing and sleeping to eating, talking and moving around – as well as all the activities of your internal organs, such as the beating of your heart, the digesting of your food, the functioning of your reproductive organs and, most important of all, the working of your brain.

## A LIFETIME OF ACTIVITY

Thyroid hormone is active when the embryo is still in the womb, where it plays a crucial part in helping each of the millions of cells in our bodies to become more specialized. It is this process, known as differentiation, that turns tadpoles into frogs and a human embryo to develop from a tiny cluster of cells into a fully grown baby.

At the other end of life, the thyroid is thought to play an equally important role in the control of ageing.

## THE GROWTH FACTOR

In the womb and after birth, thyroid hormone is vital for both mental and physical growth. With somatotrophin (STH) from the pituitary, it determines the length and strength of your bones. During childhood, lack of TH stunts growth by preventing the bones from growing and maturing. TH is also crucial for the normal development of the brain and nervous system in both the unborn and newly born infant. During pregnancy, low levels of $T_4$ can affect brain development, resulting in mild-to-severe mental deficiencies. In the past, the term used to describe these defects was 'mental cretinism'.

## BREAST DEVELOPMENT

Thyroid hormone may also be involved in the development of our breasts. Studies in mice have shown that TH affects prolactin, another pituitary hormone. In breastfeeding women, this hormone is involved in stimulating the production of milk. It is also thought that breast pain that is not premenstrual breast tenderness and swelling may be linked to thyroid problems.

## PROTECTION AGAINST STARVATION

When it works properly, the thyroid plays a crucial part in keeping your body weight more or less stable. Increasing the amount you eat, especially of starchy foods or carbohydrates, increases metabolism and boosts the production of the active thyroid hormone $T_3$. Dieting, on the other hand, decreases metabolism, causing the body to produce less $T_3$.

This is almost certainly a mechanism that has evolved to protect us from starvation. It is a known fact that when the body is deprived of food, it turns down the rate of metabolism. This is one of the mechanisms thought to have enabled the survival of the babies who, incredibly, were found still alive after several days trapped under the rubble of the Mexican earthquake in the 1980s. This same mechanism also explains why the thyroid becomes sluggish in women with eating disorders such as anorexia, bulimia and excessive dieting. The brain correctly perceives these states as starvation and turns down thyroid activity to conserve energy. This is how a thyroid problem can play havoc with your appetite and your weight.

## PROTECTION AGAINST INFECTION

The thyroid is a vital part of the body's immune-defence mechanism. It stimulates the production of special white blood cells, known as T cells and B cells, to help the body fight against disease. Chronic liver and kidney disease, acute and chronic illness, starvation and diets too low in carbohydrate lower the production of $T_3$. It is thought that this may be part of an adaptive process to help the body defend itself against illness.

FLUID BALANCE

The thyroid plays a vital role in a myriad other bodily processes. It helps to maintain the body's fluid balance by controlling the mechanisms by which water and chemicals enter and leave the cells – one reason why bloating is troublesome if you have an underactive thyroid.

VITAMIN POWER

In the liver, thyroid hormones are needed to convert beta-carotene (the pigment that gives orange, yellow and red fruits and vegetables their colour) into vitamin A. In the past few years, beta-carotene has sparked considerable interest as one of the three key antioxidant vitamins (the other two are vitamins C and E) that play a crucial role in protecting the body against degenerative diseases, such as cancer and heart disease, and those associated with ageing.

INTERACTIONS WITH OTHER HORMONES

Thyroid hormone needs to be present for other hormones to function in various parts of the body. Most important for women, it acts in concert with the female sex hormone oestrogen to modulate reproduction. This is why thyroid malfunction can sometimes be a cause of reduced fertility and other reproductive problems.

## Manufacturing Thyroid Hormone

The mineral iodine – a trace element found in soil and food – plays a central role in the manufacture of thyroid hormone. Iodine is needed for cells to work properly. To produce hormone, the thyroid absorbs iodine and, through a process involving enzymes, combines it with the amino acid tyrosine. More enzyme reactions convert this into $T_4$ and $T_3$, which are then stored by the thyroid within a protein called thyroglobulin (TG). When thyroid hormone is then needed in the body, enzymes break down this TG to release the $T_4$ and $T_3$ into the bloodstream.

## Transporting Thyroid Hormone

Most of the thyroid hormone in the body is carried around the bloodstream attached to special transport proteins, especially thyroid-binding globulin (TBG). Once the bound thyroid hormone reaches its destination, it is released from the protein binding so that $T_4$ can be converted to $T_3$ and ready for use by the cells. A tiny amount – around 0.03 per cent of $T_4$ and 0.3 per cent of $T_3$ – remains unattached to float freely about in the blood. Although only a small quantity, free-floating $T_3$ does not have to be released from any binding and so is immediately available for use by the cells.

Certain conditions, such as taking the Pill, can raise the levels of protein in the blood and, in the past, thyroid tests which measured total levels of $T_4$ and $T_3$ were not always accurate because of some confusion in interpreting results. Today's blood tests measure levels of both free-floating $T_4$ and bound $T_4$ as well as thyroid-stimulating hormone, which provide a much more accurate indication of thyroid function.

## What Can Go Wrong?

The most common thing to go wrong with the thyroid is autoimmune thyroid disease, when the body turns against its own tissues and tries to destroy them. Over a period of time, this causes the thyroid to become either overactive (hyperthyroidism) or underactive (hypothyroidism). The result is Graves' disease, the most common form of hyperthyroidism, and thyroiditis, the most common form of hypothyroidism.

But why should a mechanism designed to protect our body and keep it healthy go so drastically wrong? To find the answer, it is necessary to delve deeper into the fascinating science of immunology.

## Immune Reactions

The body's defence system normally provides a formidable barrier against attack by 'foreign' invaders such as viruses, bacteria and parasites. Although the way it works is still not fully understood, what is known is that many actions depend on two kinds of lymphocytes (white blood cells) – T cells and B cells – responsible for fending off attackers from outside.

When the body is under attack by invaders, the immune system sends T cells to the affected site to find out what is happening. There are two types of T cells: helper cells and killer cells. Helper cells help the immune system by identifying antigens, a chemical substance that marks the invaders as 'foreign'. Once the helper cells have recognized a foreign antigen, killer cells are despatched to attack and destroy them. To protect the body against future attack by the same foreign invaders, killer T cells retain a 'memory' of their antigen. If the body is threatened again by the same invader, these killer T cells are quickly activated and sent in for the kill. This entire process is known as cellular immunity.

B cells work in a similar way except that they fight off an attack by producing protein antibodies known as immunoglobulins; these are produced specific to the invader. When the immune system identifies a particular invader, B cells are stimulated to produce a large quantity of a specific immunoglobulin that will attach itself to the invading antigens and immobilize them. Once this has happened, the antigens are devoured by other white cells called phagocytes. This process is known as humoral immunity.

## The Enemy Within

Under normal circumstances, the immune system does not turn against itself because our own cells are coded to allow our T cells and B cells to recognize them as 'self' and refrain from attacking them. Scientists still do not know exactly why this normal recognition process fails. One theory is that, in some cases, a foreign antigen – say, a protein on a virus – escapes the immune system's surveillance by disguising itself as one the body's own cells. As the immune system cannot distinguish this disguised protein from its own tissue, it allows the invader access to the cells.

## Inflammation From Within

Another way in which the immune system can go awry is when an invader triggers an overzealous immune response, unleashing a flood of cell proteins called cytokines; these cause inflammation and increase the production of antibodies, which turn against the tissue in question and destroy it.

Both Graves' disease and Hashimoto's thyroiditis are caused by T and B cells that infiltrate the thyroid, triggering inflammation and the production of thyroid autoantibodies. Depending on which autoantibodies are produced, this will lead to either overproduction or underproduction of thyroid hormone.

Autoantibodies can attack virtually any of the body's tissues or organs and not just the thyroid, causing a range of diseases in which the tissues become inflamed and are gradually destroyed. These include rheumatoid arthritis, where autoantibodies destroy the joints; Addison's disease, where autoantibodies damage the adrenals; multiple sclerosis, where autoantibodies are directed against the nervous system; and diabetes, where autoantibodies turn against the pancreas.

It is now established that all of us possess autoimmunity to some degree but, in some people, the immune system seems to have a particular tendency to turn against itself. This might explain why having one autoimmune disorder can put you at an increased risk of developing another. It may also be why, if you have developed an autoimmune thyroid disorder, it is important to be on the look-out for other autoimmune problems.

Three-quarters of cases of autoimmune disease occur in women. What exactly triggers the immune system in women to attack itself more frequently than in men?

## The Family Factor

The tendency for autoimmune diseases to run in families provides one clue. It suggests that a faulty gene or genes may be partly to blame. Indeed, scientists have identified a handful of so-called 'susceptibility' genes that render some of us more vulnerable to autoimmune attack. They have also pinpointed several areas – 'susceptibility regions' – on chromosomes, the 23 pairs of rod-like structures that carry our genes, that appear to confer a greater risk of autoimmunity.

Many autoimmune diseases, including autoimmune thyroid disease, are strongly associated with a gene for human leukocyte antigens (HLA) found on chromosome 6. Another susceptibility gene, CTLA-4 (cytotoxic T lymphocyte-4), is also involved. Both these genes are known as immune-modifying genes – they alter the way in which your immune system behaves. As well as susceptibility genes, researchers are also finding genes that are specifically involved in autoimmune thyroid problems. These 'thyroid-specific genes' are thought to work hand-in-hand with susceptibility genes to trigger an autoimmune attack against the thyroid.

As to why women are at greater risk than men, scientists have come up with an intriguing theory that suggests that it may be connected with the continued presence of foreign cells from a fetus in the mother's bloodstream – and vice versa. Another way to acquire cells that aren't your own is from a twin, even one you didn't know you had, because it is now known that a number of pregnancies start out as twin pregnancies, but soon lose one of the embryos. Says Dr J. Lee Nelson, the American scientist who pioneered this theory, 'Our concept of self has to be modified a little bit. We're not as completely self as we thought we were.'

## Deciphering the Clues

The presence of thyroid autoantibodies in your bloodstream is an important clue that there has been an immune attack on your thyroid. In fact, it was by studying Graves' disease that scientists acquired some of the earliest clues of what was going on in autoimmunity. The chief culprit in Graves' disease is an antibody, first discovered in the blood of Graves' patients as long ago as 1956, dubbed 'long-active thyroid stimulator', or 'LATS', because in animals, it stimulated thyroid activity for longer than thyroid-stimulating hormone (TSH).

Later researchers identified LATS as a type of immunoglobulin G, the main antibody in the bloodstream and, because it stimulates thyroid production by locking onto the TSH receptor, they renamed it TSHR-Ab. This autoantibody is now thought to be responsible for the thyroid overactivity in Graves' disease, and to play a key role in the development of thyroid eye disease by overstimulating certain cells that line the eye sockets.

A similar process is involved in Hashimoto's thyroiditis except that, in this case, the rogue antibodies are directed against thyroglobulin (TG), the protein molecule in which thyroid hormone is stored, and thyroid peroxidase (TPO), a key enzyme involved in the early stages of manufacturing thyroid hormone. The autoantibodies block receptors on both TG and TPO, thereby causing underproduction of thyroid hormone.

# CHAPTER THREE

# The Out-of-Balance Thyroid

Given the wide-ranging action of the thyroid, it is hardly surprising that, when something goes wrong, it affects the entire body. Exactly what these effects are depends on whether your thyroid becomes underactive or overactive.

Hypothyroidism, or the underactive thyroid, can produce a long and bewildering list of symptoms (*see Table 3.4, page 42*). Most of these are non-specific and easily attributable to some other disorder or simply fatigue – one reason why it often takes so long to get a diagnosis. As Camille recalls:

> *I noticed that my mental energy had gone right down, but I kept rationalizing. The tiredness was dreadful, but I persuaded myself it was because I was overdoing it. I kept saying to myself, 'If only I'd taken two weeks off at Christmas, I wouldn't be feeling so tired'. It was only the hair loss that got me in for a test.*

Clare has a similar story:

> *I just thought I was putting on weight. I put on two-and-a-half stone in as many years. Yet, despite going to Weight Watchers and not cheating, I couldn't shift it. In retrospect, there were other clues. I developed coarse skin but,*

*because I'd had a baby, and my hands were in
and out of sterilizing solution, I just thought it was
that. My periods were irregular and I was tired all
the time, but I put that down to working and
having a family. It was sheer vanity that drove
me to the surgery in the end.*

Jennifer, who developed an underactive thyroid after the birth
of her second child, remembers:

*My energy levels fluctuated from day to day.
I would start the week feeling fine but, by
Tuesday, I would be completely exhausted and
have to take the day off. I managed to drag
myself through Wednesday and Thursday, and
Friday I had off. I would spend the weekend in
bed. I was so depressed, I would sometimes just lie
there and cry. I had constant headaches and sore
throats, my muscles ached, my nails were brittle,
and I was always getting flu. I couldn't
concentrate; my memory was appalling. I was so
cold that, even in the summer, I had to take a
hot-water bottle to bed. Our sex life went
completely downhill.*

The key characteristic of hypothyroidism is that all your
systems slow down as a result of metabolism running on
near-empty. Your appetite decreases and what you do eat is
converted into energy more slowly. You gain weight and feel
permanently cold. The smallest task becomes a supreme effort.
Your muscles feel weak and stiff, and ache on the slightest
exertion. Just walking up the road can leave you exhausted and
breathless. You may experience muscle cramps. Your heart
beats more slowly and your pulse is slowed while blood pressure rises. Digestion takes longer and you become constipated.
You may also experience joint pain and stiffness. Your kidneys
work more slowly, leading to water retention and tissue

swelling (oedema). Your liver also slows down, resulting in a rise in levels of 'bad' LDL cholesterol and other blood fats known as triglycerides. You may succumb to every passing minor infection as the lack of thyroid hormones takes its toll on your immune system. Cuts and bruises take a long time to heal because of the fragility of your blood vessels. You feel miserable, washed out and overwhelmed with fatigue. As Christine observes, 'It is total; every body system is affected. People often say, "I just feel so ill, but I can't put my finger on it."'

## Appearance Matters

One of the most distressing aspects of hypothyroidism is the effect on your looks. Even though you have no appetite, the weight piles on unstoppably. Your hair becomes dry, brittle and thin; your skin becomes dry, coarse and puffy. Your waistband nips, your rings become tight and you feel bloated. These symptoms are the result of an autoimmune attack called 'myxoedema', where the cells become 'leaky', leading to fluid accumulation and mucus deposition beneath the skin. You may become pale due to anaemia, and your complexion may take on a slightly yellowish hue due to the buildup of the yellow pigment beta-carotene in your blood.

Carol, who was initially diagnosed with depression and went for many years before her underactive thyroid was diagnosed, recalls:

> I developed nasty sores on my skin, mainly on my arms, but also on my upper thighs, stomach, and neck. It took five years to find out this was a combination of a side-effect from the antidepressants and poor skin-healing due to my thyroid problem.

You may also develop a goitre (*see page 56*) or, alternatively, your thyroid may shrivel up (atrophy).

## Tired All the Time

These physical symptoms are compounded by an almost overwhelming exhaustion, as Maggie, who developed an underactive thyroid after the birth of her first child, relates:

*I felt totally paralysed for three days. I was so weak I could barely walk around. That slowly improved, but I still felt slow – the way I imagine an old person must feel. I had no appetite, but even so, the weight piled on. On one occasion, I was actually vomiting for three days and I still put on a pound! I was freezing cold all the time and had to keep the heating turned up high. My face was puffy; I looked as though I had been crying. My head felt as if it was full of cottonwool. I couldn't focus properly – if I looked at the TV and then tried to look at a newspaper, everything was blurred. I had noises in my ears. I slept very badly. I had pain and tingling in my hands that woke me up. I was also suffering from terrible constipation. I started losing my hair, but I just thought that was the normal hair loss that happens after pregnancy, but what was strange was that I didn't have to shave my legs or pluck my eyebrows. I felt as if my whole appearance was changing. The smallest task seemed enormous – I had trouble just walking to the corner of the road. Things came to a head when we went for a walk with some friends we were visiting. I was dragging myself along at my usual snail's pace, several yards behind. Being unable to keep up with the others really brought it home to me that it was more than just the after-effects of having a baby. Something was seriously wrong.*

June said, 'I thought everyone else was going too fast. I didn't realise it was me that was slow.'

In some cases, the tiredness associated with an underactive thyroid is so troublesome that a number of doctors believe that ME/CFS (myalgic encephalomyopathy/chronic fatigue syndrome) and fibromyalgia (pain in the soft tissue and muscles accompanied by exhaustion) may be a result of undiagnosed hypothyroidism. In the absence of hard evidence from studies, there is much debate over this issue. In the US, Dr John Lowe, an expert in fibromyalgia, believes there is a clear link between the two conditions (*Clinical Bulletin of Myofascial Therapy*, 1997; see also www.thyroid.about.com). In the UK, Dr Charles Shepherd has found that hyper- and hypothyroid problems are associated with ME/CFS. Most conventional doctors tend to pooh-pooh the idea, but many current texts on the thyroid, especially those coming from America, consider the idea to have merit.

## The Senses

An underactive thyroid can affect your senses as a result of tissue swelling. You may experience headaches, migraine or blurred vision. You may become slightly deaf or hear constant noises in the ears (tinnitus). Your voice may become deep and husky due to thickening of the vocal cords. Swelling and thickening of the tissues around the wrists and ankles can compress the nerves, causing pins-and-needles in the hands and feet or numbness, a condition known as carpal–tarsal, or carpal tunnel, syndrome. This can make it difficult to use a keyboard or perform other everyday tasks. Thickening of the neck tissue may cause snoring. Digestion may be impaired as the muscular contractions (peristalsis) that propel food through the gastrointestinal tract slow down. Sluggish bowels can cause constipation.

## Menstrual Disturbances and Subfertility

Menstrual problems and subfertility – difficulty conceiving and/or maintaining a pregnancy – are both associated with an underactive thyroid (*see Chapter 8*). It may also be difficult to conceive because your sex life has ground to a halt. When doing anything is an effort, it may be impossible to summon up the energy for sex, especially if you are feeling unattractive due to various physical changes.

If hypothyroidism is a result of pituitary malfunction, you may begin to produce milk from your breasts even though you are not lactating, a result of an abnormal production of the milk-producing hormone prolactin by the pituitary.

Lyn, who was diagnosed with hypothyroidism after the birth of her second baby, says, 'I was very irritable and my lack of libido didn't help my marriage.'

## Effects on the Heart

One of the most serious consequences of hypothyroidism is that the heart, like every other system in the body, slows down. Sluggish thyroid function causes excess LDL cholesterol to accumulate in the bloodstream, which can lead to atherosclerosis – narrowing and 'furring' of the arteries – causing insufficient oxygen to reach the muscles of the heart.

Clues that your heart may be affected include shortness of breath on exertion or, sometimes, chest pain (angina). Another symptom of atherosclerosis is pain in the calf on exertion (intermittent claudication), caused by furred arteries in the leg.

Tests may reveal a slow pulse rate (under 60 beats a minute), unusual in everyone except trained athletes, low blood pressure, unusual in everyone except the very young and/or very fit, and raised levels of 'bad' LDL cholesterol and other blood fats called triglycerides.

## Mental Effects, Depression and Mood Swings

Mental sluggishness is a commonly reported effect of hypothyroidism. Your brain feels like cottonwool, and you find it difficult to pay attention, to concentrate and to remember. There may be a time lag while you try to recall events, and even familiar names or facts can be elusive – a mental state typically described as 'feeling in a fog'.

One of the biggest bones of contention is the relationship between thyroid problems, depression and mood swings. In a letter to the *British Medical Journal* (October 2000), consultant psychiatrist Martin Eales outlined his belief that faulty thyroid function – including so-called mild or subclinical thyroid problems – is a significant factor in triggering depression and the failure of some people to respond to treatment with antidepressants, and can aggravate mood swings in manic-depression (known medically as bipolar disorder). This idea receives some support from the fact that antithyroid peroxidase antibodies, associated with hypothyroidism, have been found in people with manic-depression. In rare instances, there may be more severe mental disturbances, such as paranoia (feelings of persecution). These symptoms – at one time cruelly described as 'myxoedematous madness' – quickly disappear once treatment to correct the underactive thyroid is begun.

## A Matter of Chemistry

To understand these links, it is necessary to look more closely at the chemistry of the brain and some of the discoveries that have been made concerning how the brain and body 'talk' to each other. Depression has been found to be linked to changes in both the hypothalamus–pituitary–thyroid axis and the hypothalmus–pituitary–adrenal axis – two key hormonal circuits that link the brain and the body.

A major step towards understanding depression came with the discovery that the condition is linked to a shortage of the brain chemical serotonin, sometimes called the 'happiness hormone'. This led to the development of a new class of anti-depressants called selective serotonin reuptake inhibitors (SSRIs) which, as the name suggests, selectively block serotonin receptors to cause levels of serotonin – and feelings of wellbeing – to rise. These drugs, of which Prozac is the most well known, are now considered the standard treatment for depression.

Recently, researchers found that people who are depressed tend to have raised levels of thyroxine ($T_4$). At the same time, they have disturbances in their body clock causing lower day-time levels and a lower-than-normal night-time surge of thyroid-stimulating hormone, thought to be due to lack of serotonin. This is yet another example of how the brain and body communicate. Some doctors also suggest that the activity of $T_3$ is also reduced, although this is not as yet supported by any clear evidence.

A number of psychiatrists, particularly in the US, have found that a $T_3$ plus antidepressant 'cocktail' helps lift depression faster in the 30–40 per cent of people who seem to be resistant to antidepressants. Interestingly, adding the more usual thyroid treatment, $T_4$, has not proved effective, which suggests that the ability to convert $T_4$ to $T_3$ in the brain may be damaged in depressed people. Although most studies were carried out with the older tricyclic antidepressants that are not as widely used these days, it is thought that adding a dash of $T_3$ to an SSRI might be equally effective.

## Sorting It All Out

It may not always be easy for you – or your doctor – to distinguish between depression and hypothyroidism because the two conditions have many symptoms in common. In fact, some doctors surmise that an underfunctioning thyroid may be an indicator of depression. Others think that depression can put you more at risk of developing thyroid antibodies by impairing immune function, which may, in turn, lead to hypothyroidism.

Carol's story is fairly typical:

> *I felt extremely fatigued, had trouble getting up in the morning and wanted to sleep all day. Sometimes I took a day or two off work and did just that – slept for 24 hours. I was also very tearful and had problems concentrating, making decisions, even thinking. I couldn't watch TV or read – it was too much effort. I took about three months off work. I had a nice GP at the time and he diagnosed depression. I had a feeling it was something more physical as I didn't feel depressed in the way it was described in the books. I wanted to do things, but I had no energy. I took antidepressants for five years. They helped my mood, which enabled me to return to work, but I had little energy for anything else.*

---

### Table 3.1
### Symptoms common to both
### hypothyroidism and depression

- Feeling miserable and 'down in the dumps'
- Anxiety
- Irritability
- Loss of interest or pleasure in things that you used to enjoy, such as sex
- Lack of energy
- Feeling tired all the time
- Weight changes
- Appetite changes
- Sleep disturbances
- Difficulty concentrating

## DEPRESSION OR THYROID?

---

### Table 3.2
### Clues to help you to determine whether you have hypothyroidism or depression

**Hypothyroidism**
- You gain weight despite having little appetite and eating small amounts
- You sleep more but still feel that you can't get enough sleep
- You experience low self-esteem and loss of confidence due to inability to function at home and/or work as well as changes in your appearance

**Depression**
- Weight changes are more directly linked to changes in appetite – weight loss is associated with lack of appetite and weight gain is associated with overeating
- You wake up early feeling particularly low. This is a characteristic feature of depression
- You experience low self-confidence, but feelings of guilt are more pronounced

---

Christine, whose underactive thyroid went undiagnosed for years, urges women not to be fobbed off with a diagnosis of depression if the symptoms don't improve with antidepressant treatment. She was one of the few who developed myxoedema coma, a potentially life-threatening condition in which body temperature drops severely, brought on by untreated hypothyroidism. It can also cause low blood sugar and seizures, and lead to death. The coma can be triggered by cold, illness, infection or injury, and drugs that suppress the central nervous system. Although rare, it can still happen, as Christine recalls:

*After years of to-ing and fro-ing to the doctor, I was referred to a psychiatric unit and diagnosed as chronically depressed. I was prescribed lithium*

> *[a drug used to treat manic-depression]. Within six months, I was comatose – my body grinding to a halt and my kidneys failing. I heard the doctors talking outside my room saying it should never have happened.*

Although such an occurrence is extremely rare, it does underline the importance of persistence and of getting a proper diagnosis.

## Why Does the Thyroid Become Underactive?

There are two main types of hypothyroidism: primary, when the thyroid is the source of problems; and secondary, when a fault in the hypothalamus or pituitary has a knock-on effect on the thyroid.

* **Primary hypothyroidism** can be brought on by:
* thyroiditis (inflammation of the thyroid), a feature of Hashimoto's thyroiditis (*see page 49*) and postpartum thyroiditis (PPT)
* surgical or medical treatment for an overactive thyroid (*see Chapter 5*) or surgery and/or radiotherapy for certain kinds of cancer
* prescription medications and over-the-counter drugs containing iodine (iodides), such as lithium for treating manic-depression, and some cough remedies
* congenital (inborn) problems affecting the thyroid, such as absence or abnormal development of the thyroid or errors of metabolism (*see Chapter 9*).

* **Secondary hypothyroidism** is caused by failure of the hypothalamic–pituitary–thyroid hormonal axis leading to deficient secretion of hormones by the hypothalamus or pituitary, caused by:
* known damage to the hypothalamus or pituitary as a result of previous surgery, meningitis, trauma or radiation to the brain
* the development of tumours or cysts.

## Variable Symptoms

Although many of the symptoms of an underactive thyroid are common to both primary and secondary hypothyroidism, there are some suggestive differences that either you or your doctor may notice.

---

### Table 3.3
### Clues to help you determine whether you have primary or secondary hypothyroidism

| Primary hypothyroidism | Secondary hypothyroidism |
| --- | --- |
| Periods are likely to be heavier | Periods may be absent (amenorrhoea) |
| Skin and hair are coarse | Skin and hair are dry; skin may be pale and lacking in pigment |
| Breasts remain normal-sized | |
| Heart is enlarged due to pericardial effusion (fluid accumulation in the sac surrounding the heart) | Breasts may be small and shrunken |
| | Heart is small |
| Blood pressure and blood sugar levels are normal | Blood pressure is low, often with low blood sugar (hypoglycaemia) due to adrenal insufficiency or a shortage of growth hormone |

---

## Is Your Thyroid Underactive?

Symptoms of hypothyroidism are not always easy to detect. Table 3.4 lists some symptoms that you may experience, that others may notice or that your doctor may detect.

## Table 3.4
## Symptoms suggestive of hypothyroidism

**You may experience**
Anxiety and panic
Bloating and
  fluid retention
Blurred vision/
  difficulty focusing
Brittle dry nails
Breathlessness/
  chest pain on
  exertion
Constipation
Depression
Difficulty in
  concentrating/
  conceiving/
  exercising
Discomfort
  swallowing
Dry eyes
Dry skin
Fatigue
Feeling
  constantly cold
Forgetfulness
Headaches and
  migraine
Increased infections
Joint pain and
  stiffness
Menstrual problems
Miscarriage
Muscle weakness
  or stiffness and
  cramps
Sluggish digestion
Tingling and
  numbness in hands
  and feet (carpal tunnel
  syndrome)

**Others might notice**
You are moody
You've put on weight
You have to peer at
  a menu to read it
You puff and pant
  when you walk up a
  hill
You seem miserable
You don't seem to be
  following what they
  say
You've lost your
  enthusiasm
You seem unusually
  irritable
You've gained weight
Your face and eyes
  appear puffy
You've started snoring
Your skin looks
  faintly tanned (due to
  beta-carotene
  deposited beneath the
  skin)

**Your doctor may detect**
Anaemia
Doughy abdomen
Goitre
Galactorrhoea
  (milk secretion when
  you aren't
  breastfeeding) due to
  pituitary dysfunction
Loss of muscle power
Oedema (swelling)
Slowed reflexes
Increased pigmentation
  (due to beta-carotene
  deposited beneath the
  skin)

## Life in the Fast Lane: Hyperthyroidism

Overproduction of thyroid hormones – hyperthyroidism – is caused by an overactive thyroid. The state of being hyperthyroid, called thyrotoxicosis, is sometimes easier to spot than hypothyroidism, partly because there may be rather dramatic mental and physical effects. However, symptoms are not always obvious, but may just creep up on you. It may only be when someone else comments on how you have changed or when your doctor notices some signs that the condition is diagnosed.

Whereas an underactive thyroid slows your body down, an overactive one speeds it up, causing your metabolism to race uncontrollably. As Jan, who has Graves' disease, the most common cause of hyperthyroidism, describes it:

> I'd had a lot of trouble in my marriage and a lot of stress generally after I left. The first thing I noticed was that I was full of energy. I couldn't sit still, I had to be working, working out, cooking or doing something with the kids all the time.
> I started to drink to try and slow myself down. As time went on, I couldn't sit down long enough to think and I became totally exhausted. My muscles started to waste away, even though I was exercising so much. My periods stopped. I had bouts of breathlessness, which were diagnosed as asthma. I couldn't think straight, my mind was so overactive. I felt as if my head was full of twittering sparrows and I had what I can only describe as an 'electrical buzzing' in my head.

As Jan's account illustrates, when the thyroid becomes overactive, the body burns energy at a tremendous rate. If you are affected, you can eat like an elephant without putting on weight; in fact, more often than not, you will lose pounds instead.

As the gland continues to step up production, you may feel constantly hot and sweaty, and find yourself stripping off and throwing windows open, even on cold days. You may also notice a change in bowel habits – needing to go more often and sometimes having diarrhoea, a symptom so common that it's not unusual for hyperthyroidism to be first diagnosed at a gastroenterology clinic. Some of those affected experience a raging thirst and pass large amounts of urine, similar to that seen in people with undiagnosed diabetes.

Anyone who has been around someone with an overactive thyroid can't fail to notice their boundless 'get up and go'. Sufferers pace like caged lions, talking 19 to the dozen, yet are unable to muster any concentration. Their energy never flags for a second, even at bedtime. Recalls Louise, 38:

> *I couldn't settle for the jumble of racing thoughts that were flying around my brain. My sex drive increased, too – I wouldn't leave my husband alone.*

Louise's experience echoes that of many others, and is thought to occur because of the increased turnover of male-type sex hormones – androgens – which control the libido and are converted into the female hormone – oestrogen – in the body.

## Mood Swings

Wildly swinging moods are a key feature of thyroid overactivity. One minute you are optimistic and on top of the world, the next you are plunged into the depths of despair. Nervousness and anxiety are also characteristic, probably as a result of increased sensitivity to the effects of the stress hormone adrenaline, which triggers the body's 'fight or flight' reaction. Unfortunately, some find that when they report symptoms to their doctor, they are seen to be a cause rather than an effect of their problem. Just as people with an underactive thyroid may find themselves dismissed or treated with antidepressants, it has been known for those with hyperthyroidism to be referred for psychiatric help for manic-depression.

## Appearance Matters Too

An overactive thyroid can affect your physical appearance. Your skin becomes thin, pink and moist; you tend to flush easily, and your palms may become red and sweaty. Your hair becomes fine and flyaway, and falls out while your nails become thin and flaky.

A number of those with Graves' disease develop thyroid eye disease (*see Chapter 8*) and some will also develop an infiltrating skin disorder causing the skin on the front of the shins to become lumpy, red and thickened skin in the front of the shins – a condition also known as pretibial myxoedema. These symptoms can appear years before or after the thyroid becomes overactive.

Some people with hyperthyroidism develop thyrotoxic tremor – a constant, fine trembling that is most noticeable when the hands are stretched out. This is thought to be due to an oversensitivity to adrenaline. Maria, 35, a freelance photographer, recalls that this tremor was the first thing she noticed when her thyroid became overactive:

*I first became aware of the problem when I noticed that I wasn't able to hold my camera steady. I couldn't hold a pen straight to write either, and I started having palpitations. My heart beat so fast that, on one occasion, I was convinced that I was going to have a heart attack. I was losing weight rapidly: I went from my usual eight-and-a-half to nine stone to seven-and-a-half stone, even though I was eating like a pig. And I was irritable and bad-tempered.*

## Bone, Heart and Other Muscles

Like hypothyroidism, untreated hyperthyroidism can damage the heart. The palpitations Maria describes are a common feature, caused by an overactivity of the heart muscle that causes the pulse to accelerate; this can lead to palpitations and an irregular heartbeat, called atrial fibrillation (*see Chapter 10*), especially in older women. Breathlessness is another common symptom and this, too, is sometimes misdiagnosed as asthma or bronchitis. The slightest exertion can bring on an attack.

An overactive thyroid can also disturb your body's calcium balance, accelerating the natural rate of bone loss. Bone is a living tissue that is constantly being built up and broken down. Thyroid overactivity speeds up the breakdown part of this natural cycle. This, in turn, can lead to thinning of the bones (osteopenia) and an increased risk of osteoporosis when you are older.

Weakness as a result of wasting of the muscles is another problem for about half of all hyperthyroid sufferers. As Sarah remembers:

> *I am a marathon runner, but I just couldn't run at all. If I got down on the floor, someone had to help me up.*

Very rarely – and particularly in those who are Asian – people with an overactive thyroid can experience periodic paralysis, attacks of profound muscle weakness or paralysis brought on by eating sugary or starchy foods. This is due to a disturbance of the body's ability to maintain a constant concentration of potassium in the blood.

## Menstrual Problems and Pregnancy

Hyperthyroidism, too, can be responsible for menstrual problems, including PMS. Although not as likely to affect fertility as an underactive thyroid, it is nevertheless associated with a number of complications during pregnancy (*see Chapter 9*).

## The Goitre Connection

As with an underactive thyroid, an overactive one can also cause a goitre (*see page 56*). If the doctor listens through a stethoscope, it may be possible to hear the blood surging turbulently through the vessels in the goitre, a noise known as 'thyroid bruit' (a bruit is the sound made in the heart, arteries or veins when the blood flows at an abnormal speed).

## Is Your Thyroid Overactive?

The symptoms of hyperthyroidism can often sneak up insidiously. Table 3.5 lists some of the clues that you or others may notice, or that your doctor may detect.

## Why Does the Thyroid Become Overactive?

Hyperthyroidism may also be primary, when something goes wrong with the thyroid itself, or secondary, when the fault lies with the hypothalamus or pituitary.

**Primary hyperthyroidism** can be due to:
- Graves' disease, caused by autoimmunity (*see page 50*)
- Thyroiditis (inflammation of the thyroid), caused by autoimmunity (*see page 52*)
- Nodular thyroid disease, for example, toxic multinodular goitre (Plummer's disease), characterized by the development of multiple lumps, or a 'hot' nodule (toxic adenoma), where a single lump becomes overactive (*see page 57*)
- Postpartum thyroiditis, wherein problems develop after giving birth (*see Chapter 9*)
- Excess iodine either from the diet (from food or, in some instances, herbal supplements) or from medications (such as the drug lithium, used to treat manic-depression; amiodarone, a drug used to treat irregular heart beat; and interferon, used to treat certain types of cancer)

## Table 3.5
## Symptoms suggestive of hyperthyroidism

| You may experience | Others may notice | Your doctor may detect |
|---|---|---|
| Anxiety | Agitation and | Atrial fibrillation |
| Constant hunger | nervousness | (irregular heart |
| Difficulty carrying | Argumentative | rhythm) |
| heavy objects or | Changes in your eyes | Goitre |
| standing up | Don't look as fit as | Low blood pressure |
| Dislike of heat | before | Rapid pulse |
| Frequent bowel motions | Hands are shaking | Thyroid bruit |
| Greasy skin | Moist palms | |
| Increased sex drive | Spotty face | |
| Increased sweating | Swollen neck | |
| Insomnia | Very moody | |
| Itchy, inflamed eyes | Very talkative | |
| Loss of muscle tone | Weight loss | |
| Lump in your throat | | |
| when you swallow | | |
| Menstrual problems, | | |
| such as missed | | |
| periods or cycle | | |
| changes | | |
| Mood swings | | |
| More hair in your hair | | |
| brush | | |
| Muscle weakness | | |
| Overactive mind | | |
| Problems doing up | | |
| collars or necklaces | | |
| Racing heart | | |
| Restlessness | | |
| Shortness of breath | | |
| Tendency to flush | | |
| Tremor | | |

- Overdosage of thyroxin treatment for hypothyroidism. On rare occasions, hyperthyroidism can be a consequence of people with an underactive thyroid accidentally or intentionally taking too much medication, a condition known as thyrotoxicosis factitia.

**Secondary hyperthyroidism** can be brought on by:
- Faulty pituitary function, on rare occasions due to a pituitary tumour, leading to an abnormal production of thyroid-stimulating hormone (TSH), thereby causing the thyroid to produce too much hormone
- Cancer-related problems. In extremely rare instances, hyperthyroidism may be the result of a thyroid cancer that has spread.

## Autoimmune Thyroid Problems

### Hashimoto's Thyroiditis

In adults, the most common reason for the thyroid to become underactive is autoimmunity. Hashimoto's thyroiditis (Hashimoto's disease) – named after Hakuru Hashimoto, the Japanese doctor who originally described it in 1912 – is the most common type of autoimmune hypothyroidism. The other type is called 'spontaneous atrophic hypothyroidism', where the thyroid wastes away and shrinks. This is more likely to affect older women.

THE SYMPTOMS
At first, although you may not feel ill. You may develop a small, painless goitre and, as time goes on, this may become tender and feel uncomfortable when you swallow. Curiously, when the disease first develops, you may develop symptoms of an *overactive* thyroid (*see page 48*). This is only temporary, however. As the disease progresses, the thyroid becomes increasingly less active, and the typical signs and symptoms of hypothyroidism eventually set in.

## Graves' disease

For six to eight out of 10 women, hyperthyroidism is a result of Graves' disease, another autoimmune condition that is the mirror image of Hashimoto's disease. It is most common between the ages of 20 and 40, but it can be seen in girls as young as five and, very occasionally, in the infants of sufferers.

Robert James Graves, a charismatic Irish physician, gave his name to the illness. In 1835, he wrote a paper outlining all the symptoms now recognized as Graves' disease in the UK and USA. In Europe, the same condition is often called 'von Basedow's disease', after Dr Carl A. von Basedow, a private practitioner in Germany, who described the illness in 1840. Graves was the first to make the connection with pregnancy – the women he wrote about were all pregnant (*see Chapter 9*).

## Confusing Symptoms

Graves' disease may be associated with all the classic symptoms of hyperthyroidism but, according to the UK-based endocrinologist Dr Anthony Weetman, these can be extremely variable. Writing in the *New England Journal of Medicine*, Weetman explains that both age and duration of thyroid overactivity play a part in determining which symptoms predominate. In over half those affected, nervousness, fatigue, rapid heart beat, heat intolerance and weight loss are key symptoms. However, in the over-50s, weight loss and loss of appetite are more common. Atrial fibrilliation is rare among the under-50s, but affects up to a fifth of those over 50. And while 90 per cent of younger women have a firm, diffuse goitre, only 75 per cent of the over-50s do. Glucose intolerance (inability to metabolize glucose) and, more rarely, diabetes can accompany Graves' disease, and if you have diabetes, the condition will increase your need for insulin.

## Is It Really Graves' Disease?

Graves' disease has been called the 'great masquerader' because it doesn't always produce the typical symptoms of an overactive thyroid. Confusingly, the condition can take a relapsing-remitting form in which the thyroid swings from overactivity to normal to underactivity and back to overactivity again. Even more curiously, 5 per cent of those with Graves' disease become hypothyroid over time, sometimes becoming lethargic and passive, and unable to do anything but lie in bed all day. Patricia, 34, who was diagnosed with an overactive thyroid two years ago, recalls:

*In the past, I was always a very active person. I love sports and would be out playing tennis or squash or doing aerobics four or five times a week. A couple of years ago, I began to feel completely worn out. I started to put on weight. My muscles ached all over and I felt fluey. I really struggled to get through each day. I was backwards and forwards to the doctor for about six months but, each time, I was diagnosed as having flu or a virus.*

*My mother suffers from an underactive thyroid so when my neck began to swell, I asked the doctor if I could have a thyroid problem. He said no. He thought it was a problem with my ears, because my job involves a lot of flying abroad. Eventually, I saw an ENT specialist, who felt my neck and said, 'Are you being treated for your thyroid problem?' Two days later, I was back at the hospital having tests, which showed I had an overactive thyroid. My symptoms weren't at all typical, which I guess is why it took so long to get a diagnosis.*

Such symptoms tend to be more common in older women who develop an overactive thyroid and who may be labelled depressive or thought to be suffering from a hidden cancer. This type of hyperthyroidism – known as apathetic thyrotoxicosis – can be particularly tricky to detect, which can lead to delays in diagnosis. But a diagnosis is important as this kind of apathy is a sign that the body's metabolism has reached the point of burnout and in need of urgent treatment to bring the thyroid under control.

## Thyroiditis

Thyroiditis is inflammation of the thyroid. There are three different types:

- **Viral or subacute thyroiditis** is the result of a virus infecting the thyroid, although no single virus has yet been identified as the culprit. It tends to be especially common between the ages of 20 and 50. The condition usually resolves within two to five months, although one or two out of 10 of those who get it will have a recurrence. Symptoms may include:
  - swelling, pain and tenderness of the thyroid
  - flu-like symptoms and/or a raised temperature
  - symptoms of thyroid overactivity (*see page 43*) lasting for two to four weeks, sometimes followed by symptoms of hypothyroidism.
- **Autoimmune thyroiditis** is yet another autoimmune effect on the thyroid. Mild autoimmune thyroiditis can affect as many as one in five women who are otherwise healthy, often without their even being aware of it. In a small number – about one in 10 – the disease may progress to overt hypothyroidism. The condition tends to run in families, so if you have a family history of this condition (*see page 129*), the doctor may suggest testing for thyroid antibodies.
- **Postpartum thyroiditis** (*see Chapter 9*).

## Triggers and Causes

Hashimoto's thyroiditis, Graves' disease and most kinds of thyroiditis are autoimmune conditions. What triggers the immune system to see the thyroid as its enemy in such cases? All have a genetic component, yet studies of identical twins show a relatively low genetic effect; clearly, environmental and lifestyle factors must play key roles. Research has been mainly aimed at Graves' disease, but there is reason to believe that similar mechanisms may be involved in other autoimmune types of thyroid disease.

## Could Infection Play a Part?

Because the immune system is commonly triggered by infection, the hunt has been on for some time to identify a possible specific infection that might trigger autoimmune thyroid problems. One of the most intriguing suggestions, described by US surgeon Mr David V. Feliciano in the *American Journal of Surgery* in November 1992, was that Graves' disease might be sparked off by a food-poisoning bug known as *Yersinia enterocolitica*, a distant relative of the plague bacteria.

Although *Y. enterocolitica* has not proved to be as significant as Feliciano suspected, a study published in the journal *Clinical Microbiology and Infection* in 2001 reported that patients with Hashimoto's thyroiditis also had a 14-fold increase in *Y. enterocolitica* antibodies – so the question is still open.

## Is Stress to Blame?

Over the past few years, there has been increasing evidence that, in a number of illnesses, the immune system is weakened by negative mental states such as fear, tension, overwork, anxiety and exhaustion – in a word, stress. So, could stress be responsible for autoimmune thyroid problems? The answer seems to be yes, especially in the case of Graves' disease.

Doctors in the 19th century observed that Graves' disease often followed a period of severe emotional stress – a frightening episode or 'actual or threatened separation from an individual upon whom the patient is emotionally dependent'. One 19th-century doctor, Bath-based physician Caleb Hillier Parry, described the onset of symptoms in the patient 'Elisabeth S, aged 21':

> *[She] was thrown out of a wheelchair in coming*
> *fast down hill, 28th April last, and very much*
> *frightened, though not much hurt. From this time*
> *she has been subject to palpitation of the heart,*
> *and various nervous affections. About a fortnight*
> *after this period she began to observe a swelling of*
> *the thyroid gland.*

Today, Dr Mark Vanderpump, secretary of the doctors' organization the British Thyroid Association, observes:

> *When compared with people without thyroid*
> *disease or patients with toxic nodular goitres,*
> *patients with Grave's more often give a history*
> *of psychological stress before the onset of*
> *hyperthyroidism through immune suppression*
> *followed by immunological hyperactivity. The*
> *same phenomenon is seen post pregnancy as well*
> *when the immune system is suppressed during*
> *pregnancy and relapse follows delivery.*

In Norway and Denmark, the incidence of hyperthyroidism increased during the first years of the Second World War. In their book *Thyroid Disease: The Facts*, Drs R.I.S. Bayliss and W.M.G. Tunbridge mention research showing a significant rise in the incidence of Graves' disease in Northern Ireland since the start of political troubles in 1968. More recently, researchers reported a dramatic fivefold increase of Graves' disease in eastern Serbia during the war in the former Yugoslavia.

## Significance of Life Events

Back in 1991, a team of Swedish researchers found that those developing Graves' had often suffered an unhappy event in the recent past. The death of a close relative or friend was reported by 15 per cent of Graves' patients compared with 10 per cent of a control group. The disease was also more likely to strike those who were divorced or less happy with their jobs – suggesting that long-term anxiety, unhappiness and other negative feelings could be a factor.

In 1998, Japanese researchers reported in the journal *Psychosomatic Medicine* that women diagnosed with Graves' disease were seven-and-a-half times more likely to have experienced stressful life events although, curiously, the same finding did not hold true for men. In 2001, another study, reported in the journal *Clinical Endocrinology*, found a five-and-a-half-fold increase in 'life events' in individuals with Graves' compared with those with toxic nodular goitre and those without thyroid problems. Intriguingly, in this study, people with Graves' had also experienced more happy – but still potentially stressful – events like a promotion, a pay rise, getting engaged or married, or having a baby.

None of these studies prove conclusively that stress is to blame for triggering Graves' disease, but they do suggest a significant connection. However, fewer connections have been found linking stress with Hashimoto's thyroiditis. Clearly, more research is needed to unravel the precise mechanism by which stress may tip the thyroid into overactivity and to determine whether it is a factor in thyroid underactivity.

## Smoking

Smoking has long been a known risk factor for the thyroid eye disease Graves' ophthalmology (*see Chapter 8*). However, its role in Graves' disease has been less clear. Nevertheless, the evidence is beginning to stack up. In the 1998 Japanese study mentioned above, smoking was found to be an independent risk factor for women developing Graves'. This finding was echoed in a Danish study, published in the journal *Thyroid* in January 2002, in which 45 per cent of women diagnosed with Graves' disease were current or former smokers, compared with 28 per cent of those with toxic nodular goitre and 23 per cent of those with autoimmune hypothyroidism.

## Lumps, Bumps and Swellings: Goitre, Thyroid Nodules and Cancer

Both hyperthyroidism and hypothyroidism can lead to a goitre (a swollen thyroid). As a symptom, a goitre may mean anything or nothing. Small but nonetheless visible goitres are found in about 15 per cent of the population and affect four times as many women as men. They can vary from a slight fullness in the neck to a swelling so pronounced that fastening close-fitting necklaces or a choker becomes a problem and the top button on shirts or blouses can no longer be done up.

### Causes of Goitre

The most common type of goitre is a smooth soft swelling called a simple, non-toxic goitre that doesn't affect thyroid function. In the past, iodine deficiency would have been the most common reason for goitre, and it is still the most common cause worldwide. Goitre may also be caused by autoimmune thyroid disease, where the thyroid strives to produce normal levels of thyroid hormone in the face of autoimmune attack, or by prescription drugs such as antithyroid drugs or amiodarone, taken to correct an irregular heartbeat, as well as iodine-containing over-the-counter and healthfood supplements such as kelp. Thyroid inflammation (thyroiditis) due to a viral infection (for example, de Quervain's thyroiditis) is another cause of a goitre. The whole of the thyroid may swell or it may be confined to an isolated clump of cells known as a nodular goitre.

## Thyroid Nodules

With the increased use of ultrasound scanning, it is now recognized that many people have thyroid nodules – a small knot of cells that, like a goitre, is usually of no significance. Nodules can be solid or cystic (filled with fluid) and some may be malignant (cancerous). One out of 10 solid thyroid nodules is malignant, and cystic nodules are even less likely to be cancerous. Nodules can also be a mixture of cystic and solid – known as complex nodules.

A single, or solitary, nodule may get bigger, shrink or even vanish, although most stay more or less the same. When multiple nodules cluster together, causing enlargement of the thyroid, this is known as a multinodular goitre. In this past, this was called Plummer's disease; it tends to be especially common in older women. Over time, many of these goitres grow, becoming more nodular and sometimes increasingly more active until it tips over into overt hyperthyroidism. Around five in every 100 people become hyperthyroid each year as a result of a multinodular goitre.

### HOT AND COLD NODULES

Nodules are classed as 'hot' or 'cold' according to how they appear on a radioactive scan (*see page 77*). Hot nodules absorb radioactive iodine, so they are active, functioning thyroid cells. As they are nearly always benign (non-cancerous), they are also known as toxic adenomas – an adenoma being another name for a benign lump – although they can produce excessive amounts of thyroid hormone. Hot nodules are more common in middle age and later life, and often are the result of mutated genes. This kind of gene mutation is usually not passed down through the family line and do not affect every cell in the body, but is caused by an acquired fault in the genes that control the activity of the cells in the goitre.

Cold nodules don't take up radioactive iodine and are, therefore, non-functioning. Most cold nodules are also benign, but they can sometimes be more sinister. If you have a cold nodule, the doctor may want to perform a fine-needle biopsy, in which a few cells are taken up, using a tiny-bore syringe, to check whether they are cancerous. Other tests such as a thyroid-function test and ultrasound scan and/or a radioisotope scan may also be done (*see Chapter 4*).

Treatment of benign nodules may include antithyroid drugs, surgery and radioiodine (*see Chapter 5*).

## Thyroid Cancer

In a small number of cases, a thyroid nodule may be cancerous. Thyroid cancer is one of the rarest cancers, responsible for just 1 per cent of all cancers. Its precise causes are still a mystery, but radiation exposure is a known risk factor. Individuals who have had radiotherapy treatment to the neck as children are at a higher risk because the thyroid concentrates radioactivity. Exposure to nuclear fallout is another cause, as evidenced by the high incidence of thyroid cancer among those who, as children, lived close to the radioactive fallout from Chernobyl. Radioactive iodine is only harmful to developing thyroid cells, not mature ones. Indeed, in Chapter 5, you will see that radioactive iodine is actually an extremely safe and successful treatment for certain thyroid conditions.

**Could It Be Cancer?**

Thyroid cancer is often discovered by chance. However, the following symptoms may be suggestive. If you experience any of them, make an appointment to see the doctor:

- A painless lump in the neck that gradually increases in size
- Difficulty swallowing (dysphagia)
- Difficulty breathing (dyspnoea)
- Hoarseness.

Note: Hyperthyroidism and hypothyroidism are rarely symptoms of thyroid cancer as cancer cells don't usually affect production of hormones by the thyroid.

## TYPES OF THYROID CANCER

- **Papillary cell cancer.** This is the most common type of thyroid cancer, accounting for about six out of 10 cases of thyroid cancer. It mainly affects younger people and women in particular. Papillary cell cancer is not an aggressive cancer and the likelihood of total remission is extremely high. It is thought to be hereditary in around 3 per cent of cases.
- **Follicular cell cancer.** This type of cancer is less common (only about one in 100 thyroid tumours are follicular). On the whole, it tends to strike older people and to be more aggressive.
- **Mixed papillary–follicular cell cancer.** Papillary and follicular cell cancers often occur together, but the outlook is extremely good as many are completely cured.
- **Medullary cell cancer.** This rare type of thyroid cancer sometimes, but not always, runs in families. It is caused by

the presence of a kind of cell known as a C cell, which produces calcitonin, a hormone that helps to control calcium and phosphate levels in the blood. Medullary cancer tends to be associated with abnormalities of other glands or the bones, and is one of a number of cancers caused by a single mutated gene that can be screened for by a gene test. If you have reason to suspect you come from an at-risk family, for example, because relatives have had this type of cancer, you can request a gene screen to check for the presence of the mutant gene. If you are found to have the gene, you can eliminate your risk of thyroid cancer by undergoing a thyroidectomy – having your thyroid surgically removed (*see Chapter 5*). Although this may seem a drastic step if you are healthy, most of those who have had it done consider it worthwhile as it avoids the constant worry that you might develop thyroid cancer.

- **Anaplastic thyroid cancer.** This extremely rare, but more severe, type of thyroid cancer is more common in women aged over 60; it tends to grow and spread more rapidly than other types of thyroid cancer. As yet, the outlook for this type of cancer is still poorer than that for other types of thyroid cancer.

## A Diagnosis of Cancer

Although most types of thyroid cancer are effectively treated, there's no doubt that a cancer diagnosis can be a terrible blow that can take some time to come to terms with. Because thyroid cancer is relatively rare, it is important to be treated in a specialist unit where the medical staff is familiar with this type of cancer. Such a unit should also have a specialist nurse, who will be able to give you information and support throughout your cancer experience. You may also find it helpful to contact one of the cancer charities listed at the end of this book (*see page 233*).

# CHAPTER FOUR

# Getting a Diagnosis

One of the most frustrating experiences for many women with thyroid problems is getting a diagnosis. Often, women with hypothyroidism have a particularly hard time as the classic symptoms, such as mood swings, weight gain and tiredness, are easily attributable to other factors, such as 'stress' or an inability to cope with life. As Lyn Mynott, chair of the patient organization Thyroid UK, observes:

*I receive hundreds of calls and letters from women who say that they were not diagnosed immediately – in some cases, it was years before thyroid disease was diagnosed.*

Lyn herself had a real struggle to get a diagnosis:

*I was ill with various symptoms for a very long time. The main problem was joint and muscle pain all over my body. Other symptoms were panic attacks, vertigo and 'women's problems'. The doctor sent me for test after test, but nothing showed up except that I was a hypochondriac. I put up with the pain, going to osteopaths and so on to no avail. I then had a prolapse of the womb, which we all hoped was the cause. I had an operation for this as I wanted another baby. I gave up work and managed with the symptoms*

*until I got pregnant. Strangely enough, I felt*
*better during my pregnancy but, when the baby*
*was seven months old, all my old problems came*
*back. I saw a chiropractor, who seemed to help for*
*a while, but had to give up due to money*
*problems. My GP at this time was very helpful,*
*but wanted to give me antidepressants. I informed*
*him I was only depressed due to the pain and*
*tiredness.*

It wasn't until she woke up with a lump in her neck some 12 years after the onset of symptoms that Lyn's thyroid problem was finally diagnosed.

It is well known that women are more likely than men to visit the doctor with somewhat vague complaints that can get them labelled as 'heartsink' patients. A salutary article entitled, 'Me, the heartsink patient' appeared in the *British Medical Journal* in 2000 in which British GP Sarah Evans describes how she developed thyroid disease following the birth of her first child:

*I became a remarkably regular attender at my*
*doctor's surgery for a variety of minor complaints.*
*My vague 'acopia' [lack of coping ability], the*
*ebbing of my confidence, my gradually increasing*
*bulk, and my mad professor hair all became part*
*of me. Patients could hear the cogs clicking as I*
*groped towards decisions. Family and friends*
*knew all about my struggles with domestic*
*management and parenting, my daytime naps and*
*early nights.*

Despite being a doctor herself, Sarah didn't think to ask for a thyroid check even though she had a 'strong family history'. When she finally had a blood test done, she was found to have hypothyroidism. She concludes, 'My spell of tiredness has led me to investigate and investigate my heartsink patients.'

Thankfully, there is a greater awareness of thyroid problems among GPs today. Nevertheless, countless sufferers still describe how their complaints are met with indifference, or put down to adolescent angst, postnatal depression, the menopause or just getting old.

June, 44, a university lecturer whose underactive thyroid wasn't diagnosed for more than a decade, echoes the experience of many women:

> *Even though my mother, my father and my sister all have thyroid problems, it was really difficult to get a definite diagnosis. I was sent to a gynaecologist for my period problems, a gastroenterologist for my irritable bowel symptoms, and an endocrinologist who acknowledged I was hypothyroid, but didn't think it severe enough to treat. I spent years and huge amounts of money on complementary treatments before I got any proper medical treatment.*

## 'What Can You Expect at Your Time of Life?'

Angela, another woman whose hypothyroidism took many years to be diagnosed, relates how, as a teenager, one doctor said her symptoms were due to overeating and, later, another one put her symptoms down to a failure to cope with the demands of motherhood:

> *I was overweight from the age of seven or eight, and I was always on and off diets. Just before my seventeenth birthday, my weight shot up to 11 stone (I'm 5 foot 4 inches) and, despite dieting, it wouldn't shift. My parents were so worried, they took me to the doctor. I saw the letter he wrote to the consultant. It read, 'This girl eats too much'.*

*When they did tests, they found I was hypothyroid. I was on thyroxine for years, but then they just took me off it without any explanation and discharged me. My weight continued to yo-yo, but I was fine until after I had my second baby – the pounds really piled on. Despite the fact that I kept having diarrhoea and I was breastfeeding, I got fatter and fatter. Although at the time I could only see my problem in terms of weight, looking back, there were other symptoms of an underactive thyroid: my skin was like elephant hide, and if I looked at the wall, it looked as if someone was pushing it out at me. It was very disturbing. I was so tired, too, and my short-term memory went to pot. I kept picking up the phone to call someone and then was not able to remember who I'd dialled.*

*When I went to the doctor, he put it down to the fact that I wasn't coping with two children. I remember his words: 'Most tiredness is socioeconomic'. It was only when I looked in a medical dictionary and realized that my symptoms fitted those of hypothyroidism that I went back and demanded a thyroid test.*

Sadly, the response of Angela's doctor is not untypical:

*After my first baby was born, I had what was diagnosed as postnatal depression. After the second one, I had a longer period of depression for which I was prescribed antidepressants. When I went to the doctor, the GP said, 'I'm sick of seeing you.' When I explained my symptoms, he said, 'It's nerves. I'll put you on a sedative.'*

Matters are complicated by the fact that first, as we've seen, the symptoms of thyroid problems are non-specific and that, second, the thyroid can be failing subtly over many years before it becomes apparent on the conventional blood tests for measuring thyroid function. June comments:

> *I feel that doctors, including most endocrinologists, need to be better educated about thyroid conditions and need to think more holistically about diagnosis and treatment. There is too much reliance on thyroid function tests at the expense of consideration of symptoms and how the patient feels.*

## 'I Couldn't Sit Still'

Hyperthyroid problems tend to be more quickly diagnosed perhaps because, often, the person with symptoms is so obviously edgy and hyperactive. However, the diagnosis may be missed if you are older when, as we've seen, there may not be the more obvious changes seen in younger women. Fortunately, Judith's doctor was alert to her symptoms of an overactive thyroid even though she herself attributed them to the start of the menopause:

> *I was feeling hot all the time, shaky and faint. I thought they were hot flushes and that I was going through the menopause. I was quite shocked when the doctor said I had a classic case of overactive thyroid.*

Donna, another woman with hyperthyroidism, described her experience in the newsletter of the British Thyroid Foundation:

> *I had quite severe hyperthyroidism intermittently for a very long time. I was labelled with 'personality disorder' and prescribed endless Valium for 10 years. After the birth of my last baby in 1990, we moved house and my illness came to a head. I went to my*

*new GP with dreadful fear and anxiety symptoms,*
*and had lost nearly three stone in weight in as many*
*months. He sent me home and very promptly sent a*
*psychiatrist to me, who took about 10 minutes to*
*make a diagnosis that was later confirmed by blood*
*tests.*

## If at First You Don't Succeed

The lesson to learn is that, whatever your symptoms, you must be prepared to persevere. Many doctors are the first to admit that the symptoms of thyroid disease can be confusing. If you start having hot flushes in your 40s or 50s as Judith did, it is perfectly reasonable to suspect the start of the menopause. Similarly, if you feel tired, weepy and emotional after having a baby or a miscarriage, it will not altogether surprising if it's attributed to postnatal depression.

But whatever the cause of your symptoms, you deserve to be listened to. If you feel you are having problems making yourself heard, it may help you gain confidence if you take a relative or friend with you to the surgery. Most doctors will have no objection to this although, as a courtesy, it is better to ask first.

A diagnosis of thyroid disease should take into account both your symptoms – what doctors call the clinical features – and the results of a thyroid function test. One of the main complaints among women was that doctors place too much reliance on blood test results at the expense of the clinical picture. In fact, biochemist Dr Denis O'Reilly, writing in the *British Medical Journal* in 2000, observed that 'the clinical features of hypothyroidism seem to have been relegated to the status of historical curiosities' while the clinical aspects of hyperthyroidism 'have been downgraded' to little more than lists. O'Reilly cites the *Oxford Textbook of Medicine* where, he says, the clinical diagnosis of thyroid dysfunction 'is dismissed in less than a line'. But while doctors like O'Reilly think that too many blood tests are performed, other thyroid experts argue that not enough are done, especially in early pregnancy (*see Chapter 9*).

## Describing Your Symptoms

As the average GP allocates just 10 minutes for each consultation, when making an appointment to discuss your problems, it may be a good idea to book a double session so that you can be sure there will be enough time to discuss everything. When you are consulting your doctor, tell him exactly what you have been experiencing and how it has been affecting your life. Everything you say will build up a clinical picture that helps with the diagnosis, so be sure not to leave out any symptoms, however unconnected it may seem. You may not think that talking about body hair is appropriate at a doctor's surgery, but if you have noticed that you don't need to remove underarm hair as often or, conversely, that you seem to be growing a beard, it may be a sign of thyroid problems. It may be helpful for you to write your symptoms down before the consultation, putting the most troublesome at the top of the list to make sure they have priority.

Be specific. To say 'I feel cold all the time' is too vague. Telling the doctor that, at the height of summer, 'I have to keep the central heating on and switch on the electric blanket before I go to bed' will make it clear that something is wrong. It may take just one such symptom to alert the doctor to the possible cause of your condition. In Jennifer's case, it was forgetfulness:

> *I'd been going to the doctor constantly with one
> thing and another. I kept getting sore throats,
> headaches, and felt tired and weepy, but I was
> always sent away and being told that it was a
> virus. I felt that they felt I was just being neurotic.
> One day, I got so desperate that I wrote
> everything down, and I went in and said, 'I don't
> want you to say anything, I just want you to*

*listen. My symptoms may seem trivial, but I've been experiencing them for a long time now.' The funny thing was that, when I mentioned loss of memory, the doctor immediately suggested a thyroid test.*

If you suspect you have a thyroid problem, don't be afraid to explain your suspicions and the reasons for them, such as a family history of thyroid problems or other autoimmune disease. Again, what you say may be the trigger your doctor needs to make a vital connection.

## Partnership and Dialogue

There's often the feeling that it is the doctor's job to diagnose and the patient's job simply to describe symptoms, but such an approach helps neither you nor your doctor. Effective diagnosis and treatment depends on an open dialogue between you and your GP. If you persistently feel you can't speak openly and honestly or that your doctor is not hearing you, there may be a communication problem between you and it may be time to change practices or seek the help of another practitioner.

Be prepared to supply details of the following:

- **Your medical history**. Include any previous illnesses, any previous thyroid problems, any family history of thyroid problems and any previous personal or family history of other autoimmune diseases, such as diabetes or pernicious anaemia, as these can sometimes suggest thyroid disease.
- **Your lifestyle and environment.** Include details of where you lived in the past (the soil in some areas lacks iodine and people who have lived there are more prone to iodine deficiency) as well as your work and your usual personality (thyroid problems often lead to rather dramatic changes, so if you are normally vigorous and active and have become slow and sluggish, the doctor needs to know). Tell the doctor what job you do and what

it involves. Thyroid problems can often affect work performance. One woman, who is a teacher, recalled how she always missed a particular class because she was so tired.

- **Your eating, sleeping and activity habits.** If you are still piling on weight despite dieting or you can't get enough to eat yet are rake-thin, this could help the doctor reach a diagnosis. If you used to be an active, sporty person, but now find it hard to climb the stairs, let alone pick up a tennis racket, or if you used to be a night owl, but now have to get to bed by nine o'clock, then these, too, could be important clues.
- **Other relevant information.** If you are trying for a baby, think you may be pregnant or have just had a baby and the doctor doesn't know, you should say so. Your doctor also needs to know if you are a smoker, as smoking is linked to certain kinds of thyroid problems. Make sure you tell the doctor, too, about any treatment you are receiving or medicines you are taking. Don't forget to mention over-the-counter drugs, herbal remedies, food supplements, the contraceptive pill and hormone replacement therapy as these can sometimes affect symptoms and blood tests or interact with any drugs the doctor may prescribe.

## The Doctor's Observations

Once you've described your symptoms, it's your doctor's turn to ask some questions to build up a more complete picture. At the same time, he should be paying attention to your appearance – your weight, the condition of your skin and hair, whether or not your thyroid is enlarged, the appearance of your eyes, even the way you speak and conduct yourself. A hoarse, croaky voice can be a sign of an underactive thyroid while an anxious, rapid demeanour can be a sign of overactivity.

He should perform a physical examination to confirm any observations, feeling your neck to check whether or not your thyroid is enlarged and perhaps measuring its circumference to check for a goitre. He should take your pulse and measure your blood pressure. If your eyes are affected, the doctor may use a special instrument called an exophthalmometer to ascertain how far they are protruding.

Every consultation should be a two-way street. Before you visit the doctor, decide exactly what you want out of the consultation and write down any questions. You might find it helpful to take a notebook to write down what the doctor says or even tape the consultation although, again, it is better to first check that the doctor has no objections.

## What Happens Next?

If the doctor suspects that you have a thyroid problem, he should arrange for a thyroid function test. This is a blood test designed to measure levels of the various hormones produced by the thyroid and pituitary glands to ascertain how well your thyroid is working. Sometimes the doctor will also include an autoantibody test.

The most important measurements are thyroid-stimulating hormone (TSH) and free-floating (not bound to transport proteins) thyroid hormones $T_4$ and $T_3$ ($FT_4$ and $FT_3$) in the blood. TSH is the pituitary-produced hormone that stimulates your thyroid into action. Its levels rise when the thyroid is failing and fall when the thyroid is overactive, so abnormal levels can be an early clue that something is wrong. In the past, total thyroid hormone ($TT_4$ and $TT_3$) levels were often measured, which includes the hormone bound to proteins. This is now known to be a less relevant measurement.

The basic blood test can be performed there and then at the GP's surgery. The test will be sent off to a lab and may give a clear enough result to enable your doctor to diagnose your problem. In other cases, you may be referred to the hospital for further investigations.

## Thyroid Function Tests

Thyroid function tests have improved over the past few years and are generally considered to be both sensitive and accurate. One thing you need to be aware of, however, is that different labs use different 'reference intervals' – the range considered to be normal. The precise reference range may be shown on the lab form. If it isn't, or the doctor hasn't told you, it is important to find out what range was used in order to interpret your test results.

Yet, many thyroid self-help groups complain that levels of thyroid hormone are highly individual and that, without a baseline measurement taken when your thyroid was healthy, it is difficult to know what is abnormal for you. As Lyn Mynott, of Thyroid UK, observes:

> *Your results could be anywhere within the range and you would be classed as 'normal'. If you are at the very edge of the range at either the bottom or the top, you could be classed as 'borderline'. Neither you nor your doctor truly knows what your normal is. I believe that a set of blood tests done at an early age, say before pregnancy, would at least tell us at a later stage that our levels have changed.*

Many doctors and biochemists argue that the term normal is a misnomer because it is based on the levels found in a segment of the population without overt thyroid disease. Given the acknowledged prevalence of subclinical or 'silent' thyroid problems, they argue, it's possible that the so-called normal ranges aren't any such thing and that the range should be changed to reflect this.

Another difficulty is that thyroid hormone levels are subject to many different influences. Pregnancy, the Pill and other drugs can affect thyroid output, as can other illnesses and medical conditions. The result can be that levels of $T_4$ and $T_3$ are affected, but no one really knows how. This said, the majority of experts consider thyroid-stimulating hormone (TSH) to be a highly sensitive test for primary hypothyroidism, given that TSH levels start to rise long before $T_4$ and $T_3$ fall below the lower limits and, very often, even before any symptoms are experienced. Likewise, TSH tests are able to detect suppression of TSH before thyroid hormone levels markedly drop in hyperthyroidism.

## Tests Are a Tool

Although tests are not infallible, to abandon them completely would be to throw out the baby with the bath water. Essentially, what your doctor is – or should be – looking for is the balance between the hormones produced by the pituitary and the thyroid. Ideally, the doctor should take into consideration the various test measurements as well as your description of how you are feeling – your clinical symptoms – to reach a diagnosis.

So what are these test range figures? According to Dr Mark Vanderpump, secretary of the British Thyroid Association, the following ranges for each hormone would be considered reasonable:

- Thyroid-stimulating hormone (TSH)        0.5–5 mU/L
- Free $T_4$ (F$T_4$)                       10–20 pmol/L
- Free $T_3$ (F$T_3$)                        3–7 pmol/L

The exact figures used vary from lab to lab, and there is an ongoing debate for either widening or narrowing the currently accepted ranges. The test results form will usually highlight any measurements outside the norm. If it doesn't, be sure to check with your doctor.

## Interpreting Your Results

The interpretation of the results of thyroid function tests is extremely controversial, not least because doctors cannot agree amongst themselves regarding the most appropriate reference ranges. However, it is important to bear in mind that no single measurement will give the entire picture. What the lab results provide is a snapshot of what is happening at a biochemical level. So:

**If your TSH test used a normal range of 0.5–5 mU/L**
- A TSH higher than this means your thyroid is struggling to keep working.
- A TSH lower than this means your thyroid is working too hard.

**If your FT$_4$ test used a normal range of 10–20 pmol/L**
- An FT$_4$ lower than this (with raised TSH) means your thyroid is failing to produce enough hormone.
- An FT$_4$ higher than this (with lowered TSH) means your thyroid is producing too much hormone.

**If your FT$_3$ test used a normal range of 3–7 pmol/L**
- An FT$_3$ lower than this (with raised TSH) means you are hypothyroid.
- An FT$_3$ higher than this (with lowered TSH) may mean you are hyperthyroid.

In the US, doctors tend to rely heavily on TSH levels as an indicator of thyroid function. This may be a mistake because these measurements can mask thyroid problems due to other causes, such as failure of pituitary function. As British thyroid expert Professor Pat Kendall Taylor points out in the *British Medical Journal*:

> *The practice of using results of TSH tests alone to indicate hyperthyroidism is to be deplored and has*

*led to a mistaken diagnosis in several cases subse-*
*quently shown to be cases of hypopituitarism [fail-*
*ure of pituitary function].*

In the UK and Europe, the emphasis tends to be slightly less on TSH alone, with a number of experts who believe that measuring free-floating thyroid hormone in addition to thyroid-stimulating hormone gives the most useful biochemical picture. A growing body of opinion suggests that $FT_3$ levels are particularly significant: if low, it may be an indication that the body is not converting $T_4$ into the active, usable $T_3$ form; if high, it may be a sign that the receptors that enable TSH to gain access to the thyroid are blocked by antibodies.

The exact levels at which TSH should be considered significant are also a matter of debate and disagreement. One survey published in the *British Medical Journal* in 1997 found that TSH levels above 2 mU/L – in the middle of the 'normal' range – were associated with an increased risk of hypothyroidism. According to the survey, half the population (men and women) fall into this group. By this reckoning, your TSH level could be considered 'normal' even though your thyroid is beginning to fail. Some experts believe that if TSH is above 2 mU/L, there is a strong likelihood that your thyroid isn't functioning properly; however, most conventional doctors disagree.

Similarly, in hyperthyroidism, Dr Vanderpump writes:

*Some 1 per cent of the population have early*
*hyperthyroidism, a biological state that precedes*
*clear thyrotoxicosis. The first abnormality is the*
*suppression of TSH, which occurs before elevation*
*of the thyroid hormone levels and usually before*
*the onset of symptoms. The point is that around*
*one in a hundred people are in this biochemical*
*state and don't know it unless tested. The natural*
*history is not clear, but presumably a proportion,*
*possibly around 5 per cent a year, go on to develop*
*overt thyrotoxicosis [hyperthyroidism]. There is*

*also a risk of bone thinning and heart rhythm*
*abnormality developing particularly in the elderly.*
*Yet the abnormal TSH is not detectable on a*
*blood test.*

When you get your test results, it's a good idea to write them down so that you can compare them with future tests. You may have to ask your doctor for the specific figures as he may not volunteer them, but simply tell you that your results are okay. This could result in delays in treatment, as June's story reveals:

> *In my case, there was a 150 per cent drop in*
> *thyroid function according to TSH measurement,*
> *yet all my GP was concerned about was that I was*
> *still within the normal range. Actually, my TSH*
> *was 6.4, but she told me I was in the normal*
> *range, and it was not until much later when I got*
> *access to my medical records and copies of test*
> *results that I found out it wasn't.*

## 'My Results Are Normal, So Why Do I Still Feel Lousy?'

If your test results don't show any abnormalities, but you still feel under par and have symptoms or risk factors for thyroid problems, don't feel you have to put up with it. Just as TSH levels can rise or fall before any obvious symptoms are apparent, so too can thyroid symptoms develop before changes in hormone levels become evident. You deserve further investigations to uncover the cause of your problems.

Your doctor may offer – or you may want to ask for – a thyroid antibody test. The presence of autoantibodies may be an early sign of thyroid failure. Some doctors will initiate treatment if antibody levels are raised, even if thyroid function test results are apparently normal, although others believe that the presence of antibodies alone isn't enough reason to treat. It may be especially useful to have an antibody test if you suspect that thyroid problems lie behind other medical problems you may be experiencing, such as difficulty in conceiving, miscarriage or depression following childbirth.

## Thyroid Antibody Testing

The thyroid antibody test can show the presence or absence of thyroid peroxide antibodies (TPO) or thyroglobulin (TG), which are associated with autoimmune hypothyroidism, or TSH receptor-stimulating antibodies (TSHR-Ab), which are associated with autoimmune hyperthyroidism. The presence of these antibodies is an indication that your immune system has turned against your thyroid and, especially if you have a borderline raised TSH level, can predict future thyroid failure. An antibody test can identify the origin of your problems and also give your doctor an idea of how your condition is likely to progress, which may be helpful in planning the management of the condition.

### Questions to Ask Your Doctor

- What are the reference values used by the lab?
- What are my levels in relation to these values?
- In your opinion, what are the implications of my test results?
- Do I have thyroid antibodies?
- Could I have a subclinical thyroid problem?
- If so, will this be treated?
- What criteria do you use for instigating treatment?
- What additional tests might I need?
- Will I be referred to an endocrinologist?

## Are Any Other Tests Likely to Be Done?

In some circumstances, the doctor may recommend other tests to define the diagnosis more clearly or before instigating treatment. These include:

### ULTRASOUND SCAN

This involves passing a device called a probe or transducer that emits ultrasound waves over the thyroid. The transducer converts an electrical current into sound waves and converts echoes bounced back from the tissue into electrical signals. These signals are then converted into a two-dimensional image of your thyroid that can be viewed on a monitoring screen. The procedure is often used to assess thyroid nodules. Ultrasound can show whether a nodule is solid or fluid-filled, although it cannot determine whether it is benign (harmless) or malignant (cancerous). However, according to Dr Mark Vanderpump, '[It] is rarely of much value in investigation of thyroid nodules and is overrequested as an investigation by GPs as it is not appreciated how often nodules occur without causing symptoms.' It can nevertheless be a good means of checking for recurrence of thyroid cancer following surgery.

### IODINE UPTAKE SCAN

This type of scan, as the name suggests, measures how much iodine is taken up by the thyroid gland. The test is done to confirm hyperthyroidism as an overactive thyroid will take up too much iodine. The patient is given a small dose of radioactive iodine on an empty stomach. This will either become concentrated in your thyroid gland or excreted in your urine over the next few hours. The amount of iodine taken up by the gland can help distinguish a transient episode of hyperthyroidism (thyrotoxicosis), when the uptake will be low, from more permanent disturbances such as Graves' disease, when the uptake will be higher, or a multinodular goitre. This test is also performed prior to treatment with radioactive iodine, and can be used to diagnose thyroid cancer.

OTHER TYPES OF SCANS

Other types of scans, such as computed tomography (CT) and magnetic resonance imaging (MRI), may sometimes be done if you are suspected of having thyroid cancer or if you have a large multinodular goitre.

THYROID NEEDLE BIOPSY

Also called fine-needle aspiration, in this test, the doctor removes a few cells from the thyroid gland, using a fine-bore needle, and examines them under a microscope. It is a virtually painless procedure that takes just a few seconds. The test is useful for determining whether thyroid nodules are harmless or cancerous; while not perfect, it can provide a definitive diagnosis in around 75 per cent of nodules.

## When Might I Be Referred to a Specialist?

If your thyroid is underactive, the GP may be able to reach a diagnosis and prescribe treatment without referring you on. If your thyroid is overactive – or if your symptoms are more complicated or your basic diagnostic tests are unclear – then you should be referred to a consultant endocrinologist, a specialist trained and experienced in treating hormonal problems, for further assessment and/or treatment. In the UK, recognized guidelines for referral state that anyone with hyperthyroidism should be referred for a specialist opinion at diagnosis. If you have hypothyroidism, the guidelines state you can expect a referral if:

- you are under 16
- you are pregnant or have just given birth (or your newborn baby has a thyroid problem)
- your symptoms suggest that pituitary problems may be the cause of your thyroid malfunction

- there are special circumstances that may make your medical management difficult or complicated – for example, you have a heart problem or risk factors for heart disease, or you are being treated with lithium for manic–depression or amiodarone for heart rhythm abnormalities.

## Key Features of a Specialist Thyroid Unit

If you are referred to a specialist thyroid unit, what should you expect to find? According to the referral guidelines, there should be on offer:

- A core team of professionals who understand the underlying origin and development of thyroid disorders. This should include access to and support from a specialist nurse and/or patient self-help group who have an awareness of the psychological needs of people with thyroid disease
- Access to the latest thyroid function tests and 'nuclear medicine' facilities such as radioisotope scanning and radioactive iodine therapy
- Access to an experienced thyroid surgeon, and an ophthalmologist who is trained and experienced in diagnosing and treating thyroid eye problems
- A clear treatment and monitoring plan, with the risks, benefits and appropriateness of the different treatments fully explained
- Informed staff such as a specialist nurse, who is available to discuss any queries you may have after your initial consultation
- Access to a centralized thyroid disease register (preferably computerized) so that you can be regularly followed-up to see whether your disease is being controlled and the effects of treatment
- Auditing of its outcomes to ensure the quality of both the information it offers and the follow-up care it provides.

The above information is taken from a consensus statement on good practice and audit measures in the management of hypothyroidism and hyperthyroidism by a working group of the Research Unit of the Royal College of Physicians of London, the Endocrinology and Diabetes Committee of the Royal College of Physicians of London, and the Society of Endocrinology.

# CHAPTER FIVE

# Treatment Options

Once you've been diagnosed, you may feel an enormous sense of relief and look forward to getting your problems sorted out at last. The impression given in much of the literature aimed at people with thyroid problems – especially hypothyroidism – is that the treatment is perfectly straightforward. And to the extent that your thyroid function tests can be restored to within the reference intervals, there is some truth in this. Unfortunately, many women with thyroid problems discover that it's often not this simple.

For a start, there's the whole debate over where to draw the line for normal (*see Chapter 4*). Also, many women find that it takes a long time going back and forth to see the doctor until a suitable treatment regimen is found. In addition, despite treatment, a significant number of women continue to live what one woman described as 'a half-life', dogged by debilitating symptoms that have far-reaching effects on their lives. As Sue, a counsellor and trainer from Yorkshire, describes:

> *When I was first put on medication, my life improved considerably, although I never regained my former energy or interest in life. Then the dose was altered downwards as they said the levels were too high, although I still did not feel well. Since then, I have gradually lost energy, put on two stone and had all the symptoms of underactive*

*thyroid back. Repeated visits to the doctor resulted in more blood tests and being told the thyroid levels were okay and would I like to go on antidepressants. Having had depression years ago, I knew I was not depressed. I felt my thyroid was underactive although they told me otherwise. All they could say was it was my age (55), empty-nest syndrome, SAD [seasonal affective disorder] or plain depression.*

*Over the last year, I have gone downhill fast. Pre thyroid trouble, I was a keen walker, regularly doing 10 miles. Following treatment, I kept on walking but was never able to go as far. Recently, I have only managed to walk half a mile with difficulty and have to stop three or four times for a rest. I have had to cut down my work to a couple of hours a week. I felt my life was over and I would end up in a nursing home as I could no longer look after myself since everything was such an effort. Luckily, my husband has been able to help me, but it has affected our marriage.*

Sadly, Sue is far from being alone. Symptoms of depression, tiredness and mental fog seem to be particularly obstinate even when other symptoms improve. To compound matters, many women with continuing problems find that their doctor is either not interested or sometimes downright unsympathetic.

The truth is – contrary to conventional wisdom – thyroid problems are not always simple to treat. The good news is that a number of endocrinologists are beginning to rise up to the challenge of helping those with difficult-to-control symptoms, and some are experimenting with new treatment regimens that may be better at controlling recalcitrant symptoms. However, doctors on the whole tend to be cautious and many are unwilling to prescribe treatments which they believe are not supported by enough evidence of efficacy. This means that it may be some time before new ways of managing thyroid problems become widely available so, in the meantime, you may have to persevere to find the treatment that best suits you.

This chapter covers the conventional medical thyroid treatments, some of the new thinking about treatment and some of the difficulties you may face. If you're lucky, you will have a sympathetic doctor who listens and is willing to carry on trying different ways of managing your condition until one is found that brings your symptoms satisfactorily under control. But if you are unlucky, you may have to shop around until you find a doctor willing to work with you to find the optimal treatment.

## Treating Hypothyroidism

Doctors have been searching for the most effective treatment for hypothyroidism since 1892, when extracts of animal thyroid first began to be used as a treatment. These extracts – usually fried, minced and served on bread! – contained two thyroid hormones: thyroxine ($T_4$) and triiodothyronine ($T_3$), the storage and active forms, respectively, of thyroid hormone. Animal extracts – by the mid-20th century made of dried, rather than minced, gland – were the only treatment on offer to those with an underactive thyroid for some 50 years. The trouble was, no two preparations were ever exactly the same so their action was somewhat hit-and-miss. In particular, the rapid absorption of $T_3$ meant that some people developed unpleasant symptoms of thyroid overactivity such as palpitations.

With the massive expansion of the pharmaceutical industry in the 1950s and 1960s, synthetic thyroid hormones manufactured in the lab became the mainstay of treatment and so they remain today. These preparations were cheap to produce and, because they could be standardized, avoided the uncertain effects associated with animal thyroid extract.

The most common (and still the standard) treatment became $T_4$, the storage form of thyroid hormone used to make

$T_3$, the active form of the hormone. $T_3$ was and still is less often prescribed, except for patients who cannot absorb $T_4$ or are unable to take tablets, as $T_3$ can be given by injection. Because it stays active for a shorter time, $T_3$ is also given to people with cardiovascular problems. However, this may be beginning to change.

## Mimicking the Thyroid

The rationale for prescribing $T_4$ has always been that it imitates what the thyroid does naturally. Ninety per cent of the hormone manufactured by the thyroid is $T_4$, and the active form $T_3$ is synthesized from $T_4$ as and when it is needed.

Back in the 1960s when $T_4$ first began to be widely used, patients were prescribed doses of 200–400 micrograms to control symptoms – these were huge amounts compared with the much lower doses usually prescribed today. With the advent of more sophisticated tests, doctors began to reduce the dosages as they came to realize that doses of just 100–150 micrograms could restore thyroid-stimulating hormone levels to normal.

However, the development of more sensitive tests also revealed that the amount of $T_4$ sufficient to normalize TSH without pushing the thyroid into overdrive could nevertheless cause changes in organs such as the liver, heart and kidney, and in the bones. Doctors began to worry about 'overtreatment' and fear that this might increase the risk of other health problems like osteoporosis. The result was a tendency to prescribe as low a dose of $T_4$ as possible – one that was only just sufficient to restore normal levels of $T_4$ and TSH in the bloodstream. Unfortunately, as some women on $T_4$ discovered, such low doses don't necessarily restore a sense of wellness, as Kathy, who developed thyroid disease following treatment for breast cancer, recalls:

> My GP put me on the standard synthetic thyroxine
> [$T_4$] and as soon as my TSH reached the limit of
> the normal range, she told me that was it. I still

*felt awful and my heart seemed to be missing
beats, which worried me, but the GP just shrugged
it off as normal.*

Clare, although generally happier with her treatment, also described experiencing continuing symptoms:

*At first, I was delighted because I felt much better.
It made me realize how awful I felt before. But
then I started to feel bad again. I went for another
blood test and they upped my dose. This went on
for some time until finally I settled on 150 mcg a
day, which seems to suit me fine. But the tablets
don't solve all your problems. For instance, I
expected the weight to roll off but, unfortunately,
it didn't – I'm still going to Weight Watchers. I do
feel much better. My energy levels are back to
normal, and I can go out and do everything I used
to do beforehand [but] I do get some joint pains
from time to time, and I wonder if that's to do
with my thyroid.*

These women are not alone. The newsletters produced by thyroid self-help groups contain numerous stories of the difficulties finding optimal treatment. Indeed, many women feel that their symptoms aren't really fully under control.

## Standard Management

$T_4$ – thyroxine sodium or levothyroxine sodium, as it is called in the medical jargon – comes in tablets of several different strengths. In the UK, these are 25, 50 and 100 micrograms, although different doses are available in the US. Several different brands are also available. You can find details of what's on offer in the UK on the British National Formulary website (www.bnf.org).

You'll usually be started on a dose of around 50–100 micrograms a day to allow your body to get used to the effects. After six weeks, you'll be offered a thyroid function test and a symptoms-check and, depending on the results, your dosage will be increased by increments of 25–50 micrograms, usually at monthly intervals, until symptoms abate and your thyroid function tests are brought within the normal range.

One factor to bear in mind is that thyroxine is slow working, so if you take a tablet on Monday, there won't be any perceptible biological effect until Friday. This means that some women can take some time before they begin to feel better. Nevertheless, there are also some who report feeling different virtually straightaway, although many doctors dismiss this as a 'placebo effect' – where a person's belief that she will get better with treatment causes her condition to improve. 'I felt as if I was on speed,' says one such patient.

The response to thyroxine is highly individual, and it's important to find the right dosage as too little $T_4$ means your thyroid will continue to be underactive, and too much will result in overactivity. Most women find their symptoms settle with a dose of between 50–200 micrograms, depending on the degree of thyroid failure. However, some women only feel better with the higher doses. Conventional endocrinologists claim that the correct dose is that which restores your thyroid function to 'normal' and relieves symptoms. According to Dr Mark Vanderpump, secretary of the British Thyroid Association, 'In most patients, this will be achieved by a dose of $T_4$ resulting in a normal or slightly raised blood level of free $T_4$, normal free $T_3$ and a normal or below normal TSH.'

Leaving aside what is 'normal', it can still take a certain amount of fiddling about to find the dosage that controls symptoms. As Sabeha, who developed hypothyroidism following the birth of her first baby, discovered:

> *Initially I was put on 50 micrograms of thyroxine.*
> *However, I was still feeling tired and the weight*
> *wasn't shifting, so the dose was put up to 75*

*microgams a couple of months later. When I went*
*back after another six months, the consultant said*
*the blood test showed my thyroid was slightly*
*overstimulated. I was experiencing a bit of chest*
*pain so she adjusted my dose, and I now take 75*
*micrograms and 50 micrograms on alternate days.*
*I feel better, but it has been a bit of a struggle, and*
*I'm still a good stone over what I was before*
*I became pregnant.*

One reason why doctors are careful to increase doses slowly is that hypothyroidism boosts cholesterol and, if your cholesterol level has been high for any length of time, you may have underlying heart disease. Starting thyroid hormone replacement at a full dose could precipitate angina (chest pain) or even a heart attack. Your age, how long you may have had thyroid problems and the likelihood of your having underlying heart disease are all factors that need to be taken into account.

If you are young and have become hypothyroid soon after treatment for hyperthyroidism, it may be decided to try a practically full dose (100 micrograms a day, adjusted upwards according to the results of blood tests). If you are older, and particularly if you have a heart problem or risk factors for heart disease, you'll be started on a much lower dose – say, 25 micrograms a day, increased by increments of 25 micrograms every two to four weeks.

## Treatment Controversies

A great deal remains to be discovered regarding the optimal treatment of thyroid disease:

- Should those who have signs of early 'subclinical' or mild, unsymptomatic thyroid problems be treated?
- What dose of $T_4$ is needed to alleviate symptoms?
- Could adding $T_3$ improve matters?
- Is animal thyroid extract more effective?

Thyroid experts admit that there is much to be learned about the best way of managing thyroid problems. In fact, according to a medical consensus statement drawn up in 1996:

> *Many aspects of the management of thyroid*
> *disease have not been subject to controlled clinical*
> *trials, yet there are established practices which*
> *have never been questioned.*

## Is More Better?

Until recently, women complaining of less-than-optimal well-being while taking thyroxine were often dismissed by their doctors or, like Clare – who was only 39 at the time – fed the line that 'you have to accept that, as you get older, you are not going to be as sprightly as you were.'

Some physicians, especially in the US, have now begun to prescribe higher doses of $T_4$ for certain patients. Many of these doctors believe that only by prescribing higher dosages will it be possible to achieve levels of TSH that they believe are necessary for optimal health and wellbeing. Drs Richard L. Shames and Karilee Halo Shames, authors of *Thyroid Power – 10 Steps to Total Health* (HarperCollins, 2001), are avid proponents of the idea that some women may need more $T_4$ than average particularly if they weigh more than 125 pounds (nine stones) and during periods of increased stress or major life changes.

British doctors are, in general, more reticent about prescribing higher doses of $T_4$, their argument being that the higher the dose, the more TSH is suppressed and the higher the level of free $T_3$. This situation, in turn, they argue, increases the risk of complications affecting the heart and bones. Nevertheless, some do agree that a number of patients only perceive an improvement in symptoms if the dose is raised to the point where the patient is 'slightly thyrotoxic'. If you feel this applies to you, it's worth talking to your doctor, who may then agree to prescribe a higher dose, provided your levels of free $T_3$ remain 'reasonable'.

## Does Brand Matter?

In the UK, you will usually be prescribed a generic (unbranded) version of thyroxine. However, just as some may find a branded preparation of aspirin suits them better than aspirin BP (the generic version), some patients may find they get better with a branded preparation of thyroxine. As one American living in the UK says:

> *American doctors advise their patients to avoid generic thyroid hormone on the grounds that it is far too unreliable. Interestingly, they've also got 88, 112 and 137 microgram dosages there which aren't available here. This means you end up breaking a 25 microgram tablet in half if you're trying to get an in-between dose, which is totally unscientific: the crumbs go everywhere. Alternatively, you take a high dose one day and a low one the next, which means you're on a rollercoaster ride.*

In the UK, there are issues surrounding the cost of medication. Patients in the US pay for their own medication, but generic prescribing is clearly cheaper for UK GPs, who are having to juggle the needs of their thyroid patients with those of others who have a whole range of different illnesses. Nevertheless, if you feel that despite adjustments in treatment, you are still not feeling well enough on a generic preparation, it may be worth discussing with your doctor whether it is possible to try a different brand.

## Finer Tuning

UK attitudes may be changing concerning these issues. At an endocrinology conference held in 2002, endocrinologists from Sheffield described research showing that some patients, although admittedly only a small minority, don't do well on a constant daily dose of the tablet strengths currently available. Some doctors, like Sabeha's, recommend 'split-dosing' – prescribing different doses on alternate days. However, this can lead to some women putting up with less-than-optimal feelings of wellbeing or to their unsupervised adjustment of their dosages. Researchers are now calling for the introduction of a new tablet strength of 37.5 micrograms which, they say, would allow dosages to be more finely tuned and provide better control for those who aren't doing well on the current treatment regimens.

## New T4/T3 Therapy: Is the Wheel Coming Full Circle?

Since the 1960s, a single hormone, thyroxine (T4), has dominated the treatment of hypothyroidism. Today, the tide of opinion may be turning with the emergence of more and more evidence that, for some women at least, a combination of thyroxine and triiodothyronine (T3) seems to be needed to improve mental function, control weight and restore zest for life.

A study reported in the *New England Journal of Medicine* in 1999 found that patients treated with a mixture of T4 and T3 were 'significantly better after treatment', especially with regard to symptoms such as depression, anxiety, irritability, fatigue and the foggy brain that affects so many women with hypothyroidism. The researchers concluded:

> *It seems clear that treatment with thyroxine plus triiodothyronine improved the quality of life for most patients.*

The findings of this and other studies could explain why women taking animal thyroid extract preparations often report feeling better (see below). Sue, quoted earlier, certainly found that a T4/T3 cocktail helped:

> *The day after starting [T4/T3], I was up at 6.30 and getting ready to go for a walk. I managed*

*three miles without stopping. I am able to do all*
*the things I never dreamt I would be well enough*
*to do. Everyone says how well I look. A friend*
*who had not seen me for a long time saw me last*
*week and said the change was remarkable. As she*
*put it, the old Sue was back and I had regained*
*my sparkle.*

Christine, who was misdiagnosed for years as depressive, also found $T_4/T_3$ more helpful than thyroxine alone. 'It gives me the wellbeing I don't get on $T_4$ alone,' she says. In contrast, Joan, who is slightly older and only diagnosed with an underactive thyroid 'by accident' while in hospital for another problem, is currently participating in a treatment trial of the new $T_4/T_3$ regimen. She says, 'I can't say I've noticed any difference.'

At an endocrinology conference held in March 2002, Dr Anthony Toft, a consultant endocrinologist at Edinburgh's Royal Infirmary, remarked that 'a significant minority of patients only achieve the desired sense of wellbeing if serum [blood] TSH is suppressed.' He drew attention to the fact that, in animals, it's been found that a combined slow-release preparation of $T_4/T_3$ may be closer to what happens in the body naturally, although currently – at least, in the UK – $T_3$ is not available in a slow-release preparation.

So, although the jury is still out, it could be that thyroid treatment is about to come full circle. But should we be rushing to embrace a treatment that, as yet, has not undergone large-scale trials? As Toft points out in a different article, while acknowledging that 'replacing the specific hormones missing when an endocrine gland fails' has a 'certain attraction', it is still too soon to advocate the widespread prescription of $T_4/T_3$. He believes that other factors still need to be assessed to make sure $T_4/T_3$ is safe, especially for the heart.

Medical writer Sara Rosenthal, author of *The Thyroid Sourcebook for Women*, also voices some concern. Writing on a US website, she warns that $T_3$ has not been adequately tested, and there are a number of questions that still need to be answered, including how it affects postmenopausal women, how it affects those with other health problems, how it affects people of different ethnic backgrounds and how it interacts with other medications. However, trials are now ongoing in the UK so, hopefully, there may soon be some definitive answers to some of these questions.

## Is There a Place for Animal Extracts?

When animal thyroid extracts were gradually replaced by synthetic thyroid hormones in the middle of the 20th century, a number of patients who had been taking them complained that the new synthetics failed to control their symptoms. One such patient was Pamela, then 68, who had been taking animal extracts as early as the 1960s. She was adamant that synthetic $T_4$ didn't work for her. With the animal extract, she maintained, she experienced 'instant brain function improvement, the fog cleared and the goitre had gone within two months'.

When natural thyroid extracts were phased out in favour of the synthetic versions, she developed headaches and her old symptoms returned, but she managed to survive on animal thyroid imported from Czechoslovakia and other countries still using the natural extracts.

Today, some self-help groups such as Thyroid UK are pressing for greater availability of animal thyroid preparations such as Armour, a $T_4/T_3$ mixture comprising dessicated pig thyroid that, in the UK, is currently only available by mail order from the US. Many of Thyroid UK's members swear this product has helped them to feel healthier and more vibrant than they have done for years.

But most conventional doctors are somewhat sceptical of these claims, attributing any beneficial effects to the placebo effect. Despite this, a number of patients are choosing to consult private doctors who are willing to treat with alternative regimens.

If you are tempted to try alternative thyroid treatment, it's important to recognize that, whether synthetic or animal, any preparation that includes $T_3$ must be used with care. $T_3$ is 10 times more biologically potent than $T_4$. That means that it can drive up your heart rate and blood pressure. While this may be not be a problem if you are young and otherwise healthy, in older women or in those with an existing or underlying heart problem, it could be dangerous.

So, if you think that $T_3$ might help you and you would like to give it a try, you need to weigh up the pros and cons very carefully, preferably in consultation with your own doctor or at least under the supervision of an informed and suitably experienced medically qualified practitioner.

## Time For a New Approach?

In the absence of any definitive answers, what this all adds up to is a need for more information on the various options, greater understanding by doctors of the frustration patients feel and more flexibility in prescribing. This is a tall order, given the limits on prescribing in public health systems such as the National Health Service, and the cautious tendencies of most members of the medical profession. However, June speaks for many disillusioned patients when she says:

> I think doctors are often too conservative in terms of dosage and need to be willing to prescribe more if this is necessary for the patient's wellbeing. They don't seem to think that someone who is 18 stone, as I have been, will need more thyroid hormone than someone who is 12 stone. There needs to be a wider range of treatment options available than at present. Where appropriate, doctors should be able and willing to prescribe alternative preparations based on animal extracts, which work better for some people, or a combination of these as well as thyroxine. There should be evaluation of whether some brands of thyroxine and $T_3$ work better than others, as there is anecdotal evidence that this is the case.

*Once someone is being treated, there needs to be recognition that TSH is not then a very useful measurement as it will be suppressed, in some cases completely, due to thyroid hormone being supplied externally. Doctors need to accept that higher levels of $T_4$ and $T_3$ than seems currently acceptable are often necessary before a hypothyroid patient reports feeling well.*

## Watchpoints

- Bear in mind that whatever treatment you use, if your thyroid gland has failed, you will have to continue treatment for the rest of your life.
- If you are severely hypothyroid or have heart disease, the doctor should refer you to hospital.
- Thyroxine speeds up metabolism, including the heart, so it's best to start gradually. If you experience chest pain while taking thyroxine, contact the doctor immediately.
- Thyroxine is usually prescribed as a single, daily dose. Before breakfast is convenient, but it doesn't really matter what time of day you choose as long as it's consistent.
- Coffee and alcohol may interfere with thyroxine absorption, so either avoid these or make sure you allow a few hours to elapse after drinking them before taking your treatment.
- Some people manage better if their doses are split or in a slow-release form. If symptoms don't improve or you find yourself experiencing 'breakthrough' symptoms, make an appointment with the doctor to discuss adjusting your dosage.
- If you don't respond to thyroxine or actually get worse, it could be that you are taking your medication incorrectly. Alternatively, there may be a mistake in diagnosis. In either case, you should check back with your doctor.

- Sometimes hypothyroidism is the result of a pituitary problem (secondary hypothyroidism). In this case, the doctor should first treat the pituitary and then your thyroid. If this is done the other way round, the symptoms may get worse because the cause of the problem is not being addressed.
- Occasionally, prolonged underactivity of the thyroid affects the functioning of the adrenal glands. In this case, the doctor may need to prescribe steroids (hormones produced by the adrenals) until they have recovered.
- Occasionally, a patient may be allergic to the binding ingredients used in the tablets. Although rare, if you have a history of allergy or intolerance, it's worth discussing this with your doctor.
- If you feel you need to adjust your medication, you should always discuss it with your doctor first.
- If the doctor puts you on medication for other health problems, check whether you need to adjust thyroxine dosage. Similarly, if you become pregnant, contact your doctor.
- You should have yearly blood checks and an appointment with the doctor to ensure that you are well and that your treatment regimen is controlling your symptoms.
- In the UK, computerized recall systems are being developed countrywide. This means you should be called up once a year to have your thyroid function and medication checked.
- If you continue to feel unwell, go back to the doctor and ask for a reassessment. It can often take a long time for treatment to take effect, but you shouldn't have to suffer in silence.
- If you don't feel better despite treatment over a reasonable period of time, it could be that you are one of those people who responds better to a different brand, dosage or preparation, so make an appointment with your doctor to discuss this.

- If you don't find your doctor sympathetic or communication has broken down between you, you may want to seek another opinion. Thyroid patient organizations may be able to provide a list or suggest the name of a sympathetic practitioner in your area.
- If you truly can't find a solution on the NHS and can afford it, you may want to consider private treatment.

## Crossreactions

Thyroxine may interact with a number of other medications, including over-the-counter drugs and nutritional supplements. Make sure your doctor is aware of any medication you are taking, especially if it includes any of those in Table 5.1.

## Table 5.1
## Medications with possible
## interactions with thyroxine (T$_4$)

| Medication | Possible effects | Action to take |
|---|---|---|
| Anticoagulants | Enhances response | Regular check-ups by your doctor |
| Iron (e.g. in nutritional supplements) | May block T$_4$ uptake | Avoid taking at same time |
| Calcium | May block T$_4$ uptake | Avoid taking at same time |
| Hormone replacement therapy (HRT) | Can increase need for T$_4$ | Have your doctor check T$_4$ dosage within three months of starting HRT |
| Insulin (used to treat diabetes) | Reduces effectiveness | Measure your blood glucose level and increase insulin if necessary |
| Propranolol (beta-blocker to treat high blood pressure, angina and heart rhythm disturbances) | Can increase propanolol uptake | Regular blood pressure checks |
| Certain antiseizure drugs to treat epilepsy (e.g. carbamazepine, phenytoin) | Lower T$_4$ levels | T$_4$ dosage may need to be increased |
| Cholestyramine (a cholesterol-lowering drug) | Reduces T$_4$ uptake | Take T$_4$ at least two hours before or four hours after |
| Tricyclic antidepressants | Risk of inducing abnormal heart rhythms | Tell the doctor prescribing the antidepressant that you take T$_4$ |

## Treatment for an Overactive Thyroid

Unlike hypothyroidism, treatment for hyperthyroid problems has long been acknowledged to be complicated, and while most women report an improvement within a month or so of beginning treatment, it can take as long as a year or more to start feeling anything like normal. Furthermore, just as with hypothyroidism, you may continue to feel vaguely unwell for some time after tests show that your thyroid has returned to normal. This is because the condition tends to come and go, and the response to treatment can be unpredictable and variable.

For this reason, in the UK, if you have an overactive thyroid, you should be referred to an endocrinologist or a doctor who specializes in thyroid problems rather than being treated by your GP. In the US and other countries where the GP is not the gateway to the specialist, you are likely to already be under the care of an endocrinologist.

### Treatment Options

Treatment will depend on the course of your disease and how you respond to treatment. Typically it could involve:

- Bringing your symptoms under control through the use of antithyroid drugs (thionamides). In a proportion of cases, this may be sufficient to restore thyroid function to normal, and can sometimes bring about long-term remission (diminution of symptoms). Once thyroid function is restored to normal and you go into remission, you may be able to come off the drugs.
- If you relapse – and, unfortunately, around half of those treated with antithyroid drugs do – you may need radioactive iodine therapy to kill the thyroid cells, or surgery to remove all or part of the gland. A side-effect of this is that you may become hypothyroid.

## Sorting It All Out

Figure 5.1 shows the various treatment options that may be offered to you.

**Figure 5.1**
**Treatment algorithm for those who have symptoms**
**of an overactive thyroid (hyperthyroidism)**

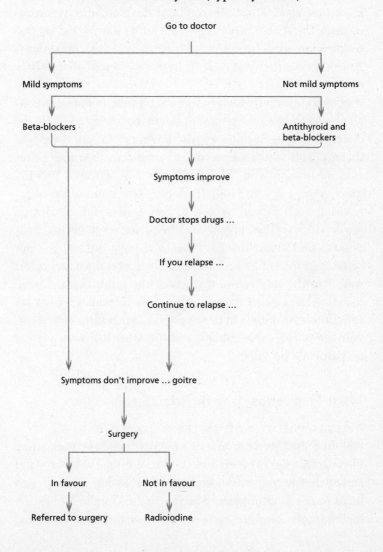

## The Medication Route

If your thyroid is only mildly overactive, or you don't like the idea of radioiodine treatment or surgery, the doctor may suggest antithyroid drugs designed to switch off the activity of your thyroid. If you are experiencing your first episode of Graves' disease, you may be offered such drug treatment for six months to two years to see if it improves your symptoms. If you are aged 40 or over, or have toxic nodular hyperthyroidism, the doctor may decide to try a treatment trial for one to three months before suggesting radioiodine or surgery. Antithyroid drugs are not effective for thyroid overactivity due to thyroiditis.

Once your thyroid has been brought back to normal – usually between one and six months after starting treatment – the doctor may suggest using what's known as 'block-and-replace' therapy. This means suppressing thyroid activity with antithyroid medication while giving additional $T_4$ (around 100–150 micrograms a day) to avoid developing symptoms of hypothyroidism. Such therapy is said to result in smoother control, fewer visits to the surgery and helps to avoid having your dosage constantly adjusted. Block-and-replace therapy is not recommended during pregnancy as $T_4$ crosses the placenta less well than the antithyroid drug, with the result that the baby may develop a goitre and an underactive thyroid. Doctors are split down the middle in their opinion of this form of therapy, with about half prescribing it and the other half saying it has no particular benefits.

## What Medication May Be Prescribed?

### ANTITHYROID DRUGS (THIONAMIDES)

An often-used medication in the UK, carbimazole (trade name Neo-Mercazole) prevents iodine uptake by the thyroid gland and so blocks the manufacture of thyroid hormone. It also helps to quell autoimmune activity. In the US, the medication methimazole (another form of carbimazole) is more often used. The biggest advantage of carbimazole is that you only have to take it once a day. A typical starting dose is 15–40 milligrams a day, reducing at four- to six-week intervals down to 5–15 milligrams a day.

An alternative medication is propylthiouracil (PTU), which is less likely to cause side-effects than carbimazole. It, too, blocks the uptake of iodine by the thyroid and prevents $T_4$ conversion into $T_3$ in certain tissues. However, PTU should not be given to anyone with obstruction of the airways or breathing passages, such as may be caused by a large goitre. Many doctors prefer to prescribe PTU in pregnancy and during breastfeeding.

## Watchpoints

- Antithyroid medication is not long-acting, but passes out of your system within eight hours. If you forget to take a dose, take double the next time.
- Antithyroid drugs can often cause itching and rashes. Although these may subside with continuing treatment, you should consult your doctor, who may be able to change your treatment. PTU seems to cause fewer side-effects than carbimazole.
- Other side-effects may include nausea, headache and, occasionally, jaundice or hair loss. Again, it's important to report such effects to your doctor.
- By far the most serious side-effect of antithyroid medication is agranulocytosis, a condition in which the number of infection-fighting white blood cells (lymphocytes) your body produces is decreased. Symptoms include high temperature, sore throat and mouth ulcers. If left untreated, it can lead to pneumonia. If you develop a sore throat or other infection, you should stop taking the medication and see your doctor for an immediate blood test. Treatment is with antibiotics to kill the infection until your white cell count returns to normal.

- Other serious side-effects include hepatitis and what doctors call lupus-like syndromes (lupus is a chronic debilitating disorder of the immune system). If you develop symptoms such as jaundice, rash, joint pain or fever, consult the doctor.
- If you've experienced any of the above serious side-effects, then antithyroid medication is not for you – you'll need to think about alternative treatment.

BETA-BLOCKERS

Beta-blockers are drugs to lower blood pressure, treat angina and regulate the heartbeat. They work by reducing some of the effects of adrenaline – the so-called beta effects. In so doing, they also reduce the amount of work the heart has to do and, thus, its need for oxygen. The anxiety experienced by most people with an overactive thyroid is a sign that adrenaline levels are raised. If your thyroid is moderately to severely overactive and you have symptoms such as a racing heart, tremor and muscular weakness, the doctor may prescribe beta-blockers as well as antithyroid drugs. Any type of beta-blocker will be effective, but most doctors prefer the slow-release types, such as propranolol and atenolol, that only have to be taken once a day. Once your antithyroid medication starts to take effect and blood tests show your thyroid function has returned to normal, the beta-blockers will gradually be tapered off.

## Watchpoints

- Beta-blockers may be the only treatment needed if your overactive thyroid is caused by thyroiditis (inflammation of the thyroid).
- Beta-blockers should not be prescribed if you have asthma, and should only be used with caution if you have signs of heart failure.
- If you can't use beta-blockers, you may simply have to make the best of it until the antithyroid medication takes

effect although, sometimes, tranquillisers may be prescribed.

## What If Drug Treatment Doesn't Work?

The main trouble with antithyroid drugs is that even if they quell thyroid activity for a time, in around 50 per cent of cases, the thyroid becomes overactive again. This is especially likely to happen if you have a large goitre or you've required large doses of drugs to bring your symptoms under control. If a relapse is going to occur, it usually does so within a year of stopping treatment. A clue that you may be relapsing is that some or all of your original symptoms return, although they may return so gradually that you don't notice until your friends or partner start to recognize signs that you are becoming hyperthyroid again. If you think you have relapsed, go to the doctor, who will arrange for further blood tests. At this point, it may be suggested that you undergo radioiodine treatment or surgery.

## Surgical Options

If drug treatment doesn't succeed in controlling your overactive thyroid, or you have a large goitre – especially one that presses on your windpipe – or a medium-sized goitre that bothers you, or you cannot or don't wish to take antithyroid drugs, or your hyperthyroidism is due to a toxic nodular goitre (see page 56), then surgery may be an option. Surgery may also be suggested if you have Graves' disease in pregnancy and can't tolerate antithyroid drugs (see page 100).

The most common procedure, in the UK at least, is a partial thyroidectomy in which, as the term suggests, part of the thyroid – usually about two-thirds – is removed. A total thyroidectomy, in which the whole gland is removed, is less widely done in the UK as it results in permanent hypothyroidism, which then requires lifelong treatment. However, there is now a school of thought that a total thyroidectomy is a preferable option.

The aim of surgery is to cure your overactive thyroid without causing you to become hypothyroid. If you do decide to opt for surgery, it is worth taking the time to find a surgeon who is experienced in performing these operations. Your doctor may suggest someone in your area or refer you to a hospital that specializes in such problems. Alternatively, you could contact one of the self-help groups for the names of surgeons known to specialize in these operations (*see page 233 for contact details*).

## How Effective Is Surgery?

Surgery restores normal thyroid function for most people with a toxic nodular goitre (*see page 56*), but if you have Graves' disease, the situation is complicated by the natural course of the disease. In some eight out of 10 cases, thyroid function is restored to normal by surgery. However, permanent hypothyroidism may develop as time goes on in an estimated 5–40 per cent. There is also the risk of recurring overactivity. This affects between one and three persons in a hundred in the first year after surgery, and about one in a hundred the year after that. These cases are usually treated with radioiodine (see below).

Factors affecting surgical success include:

- Your age
- Size of the thyroid: a small thyroid is more likely to become hypothyroid following surgery
- Size of the thyroid remnant left behind. The surgeon will usually aim to leave a remnant of less than 10 grams.
- Amount of iodine consumed, with a high level associated with recurrence.

## Before the Operation

Before the operation, your thyroid hormone levels will be restored to normal using antithyroid drugs. You will also be given Lugol's iodine – a solution of iodine and potassium iodide in water – to drink, especially if your thyroid is 'hot' (active), although many surgeons and endocrinologists today maintain that this has little or no benefit. You will be examined to make sure your vocal cords are normal. An anaesthetist will also check you over to determine the amount of general anaesthetic needed.

## The Operation

Once you are unconscious from the general anaesthesia, the doctor makes a curved incision in your neck along one of the creases to render the cut less visible after it has healed. The underlying layers of skin and muscle are carefully drawn aside to expose the thyroid; the surgeon then carefully detaches the gland from its blood supply and cuts away part of it, taking care to avoid any nearby nerves. Tubes are placed at the site of the removed gland to drain off any blood. The doctor then carefully replaces the layers of muscle and skin, and stitches or clips the incision together.

## After the Operation

Antithyroid drugs are stopped and any beta-blockers you've been prescribed are reduced gradually. Calcium blood levels are checked to make sure your parathyroid glands are working as they should. If they have become bruised, calcium levels may be temporarily low and you may need supplementation until they recover.

Usually, you need to stay in hospital for a short time – around 48 hours – but never more than five days if there are no complications. In the US, you may even be able to have the operation as a day case. As with any surgery, expect to feel some soreness at first. You may also experience some temporary hoarseness due to bruising of the vocal cords. Tubes are inserted into the thyroid for a day or so after the operation to drain off any blood, but these are removed before you go home.

After you return home, your thyroid function should be assessed about a month after your operation and then every three months for a year. After this time, you will be invited for yearly check-ups. The scar usually fades gradually until it appears as a fine line on your neck.

## Watchpoints

### BEFORE THE OPERATION

- **Clean up your diet.** Cut out alcohol and junk foods for a month before you go into hospital to reduce the stress on your liver. It may be a good idea to drink less coffee, tea, cocoa, cola and other drinks high in caffeine, as some people who consume a lot of caffeine recover more slowly from anaesthetics.
- **Vitamin power.** Fresh fruit and vegetables are rich in the vitamins and minerals your body needs to help repair itself. Vitamin C, found in all fresh fruit and vegetables, is particularly good for tissue healing. Raw foods supply instant energy and are easier for your digestive system to cope with than processed foods.
- **Don't skip meals.** Eat regularly and properly to avoid wild fluctuations in blood sugar levels, which exaggerate the ups and downs in your energy levels.
- **Relax and breathe.** Practise relaxation and avoid stress before and after your operation. Controlled breathing, such as practised in yoga, can help your body rid itself of the anaesthetic gases more quickly, and also help you have a better night's sleep – which is not always easy in hospital.
- **Pick your time.** In the book *How to Survive Medical Treatment*, medical scientist Dr Stephen Fulder suggests that women should undergo surgery during the middle of

their menstrual cycle – from days 7 to 20 – when resistance is highest, and not during a period, when the body tends to be at its lowest ebb.

- **Pack a bottle of *Arnica*.** The homoeopathic remedy *Arnica* is recommended for any type of trauma, including surgery, as it can help ease bruising and discomfort.

### AFTER THE OPERATION

- **Eat well.** Foods should be light and nourishing, yet easy to digest. Vitamin C is especially good for helping wounds to heal whereas the mineral zinc, found in seafood, liver, yeast and seeds, helps boost the immune system and aids tissue repair. B vitamins can help relieve stress, and may prevent nausea and vomiting. In their book *Superfoods*, naturopath Michael van Straten and herbalist Barbara Griggs suggest the following foods as being particularly nutritious in convalescence:
  - Fruits: apples, blackcurrants, dates, grapes, kiwi fruit, lemons, oranges, raspberries
  - Vegetables: carrots, spinach
  - Grains: oats, barley, millet
  - Seeds and nuts: almonds, chestnuts
  - Herbs: garlic.
- **Wise up to oils.** Wheatgerm oil and vitamin E, massaged over the area of the incision once it has started to heal, can help prevent scarring. Essential oils of neroli and lavender can help lift stress and ease anxiety.
- **Watch out for symptoms.** Thyroid surgery is usually extremely successful. However, you need to be aware of the possibility of recurrent hyperthyroidism or hypothyroidism and keep the lookout for potential symptoms (*see Chapter 4*). If you think you may be developing such symptoms, see the doctor.

## Are There Any Risks?

Thyroid surgery is usually safe and uncomplicated but, as with any operation, there are some potential risks. These include:

- postoperative bleeding or wound infection
- formation of lumpy scar tissue (keloids), especially if you have the type of skin that is prone to producing such scarring, most commonly if you are of African or African-Caribbean origin
- hoarseness caused by bruising of the nerve of the larynx (voice box). This affects about one in five cases and is usually temporary although, in about 25 per cent of those affected, it may be permanent. For this reason, if you use your voice in your work as a singer or teacher, for example, you may want to think carefully before opting for a surgical treatment that could leave your voice less strong and with a tendency to wobble.
- albeit rare, low calcium levels (tetany) as a result of damage to the parathyroid glands. Symptoms include numbness and tingling around the mouth, and muscle weakness and cramp in the hands and feet. In this case, long-term calcium supplementation will be necessary.

## 'I Feel Better Than I Have for Years'

Many women who have had an overactive thyroid wonder why it took them so long to have it operated upon. Angie, 48, recalls:

> Before the operation, I was really quite ill. On one occasion, I had to spend 48 hours in hospital because I kept passing out through having palpitations. The antithyroid drugs just didn't work – as fast as they gave them to me, I went overactive.

*Eventually, it was suggested that I had radioactive iodine, but I just didn't like the sound of it so the doctor suggested a thyroidectomy. I went to see the surgeon on the Friday and I was in hospital having it on the Monday.*

*Immediately afterwards, my voice got lower – which is apparently quite common – but it's back to normal now. I also noticed alarming bruising all over my chest when I first had a bath after having the operation. Apparently they have to strap you down – but no one told me and it was rather frightening.*

*I'm absolutely delighted I had it done. It's a beautifully neat job and I feel better than I have for years.*

## Radioiodine Treatment

In the US, but less so in the UK and Europe, radioactive iodine therapy (radioiodine treatment using $^{131}$I) is the treatment of choice for hyperthyroidism, especially if you are older.

Radioiodine treatment may be suggested if you have:

- problems taking antithyroid drugs for any reason
- relapsed after having drug treatment or thyroid surgery
- had a hysterectomy or been sterilized
- undertaken not to become pregnant for four months after the therapy
- a nodular goitre and you are older, your cardiovascular system is under strain from your overactive thyroid, or any other condition that makes surgery risky.

### Cautionary Tales

Although safe and effective in most instances, caution is needed if:

- you are pregnant or breastfeeding. If you are of childbearing age, you should wait four months after treatment before becoming pregnant or breastfeeding
- you have Graves' disease complicated by thyroid eye disease (*see Chapter 7*), in which case, you need to discuss your treatment with the doctor. It is better to avoid radioiodine therapy if your eye disease has been classed as active by an ophthalmologist. Occasionally, it may be better to treat for six to 24 months with an antithyroid drug to allow the eye disease to settle before initiating radioiodine treatment. In other cases, the doctor may prescribe 'prophylactic' or just-in-case steroids to prevent eye disease progression.

## What Can I Expect to Happen?

Your overactive thyroid will be brought under control by giving you an antithyroid drug (*see page 100*). This is designed to avoid the development of a thyroid 'crisis' or 'storm' brought on by a flood of thyroid hormones being released into the bloodstream as the thyroid cells are destroyed by $^{131}I$. The drug will be stopped seven days before the radioiodine is administered, as the thyroid must be working at the time of treatment, and be restarted seven days after, given the four- to six-week delay in $^{131}I$ action. If you've been on block-and-replace therapy, it should be stopped four weeks before the radioiodine treatment.

Treatment usually consists of taking a capsule containing a dose of radioactive $^{131}I$. This is taken up by the thyroid, and the radiation destroys the cells that are overactive. Over the course of four to six weeks, this calms down the overactivity of your thyroid. The treatment is almost always painless, although a small number of patients – estimated to be less than 1 per cent – may develop a temporary inflammation of the thyroid, which may be quite tender for a few days.

The doctor aims to give you sufficient radioiodine to bring thyroid activity back to normal within two or three months.

As with surgery, there is a significant risk of hypothyroidism – typically affecting 15–20 per cent of women two years after treatment and 1–3 per cent each year after that. The ideal is to use just enough radioactive iodine to normalize thyroid activity, but not so little that the gland becomes overactive again or so much that you risk hypothyroidism. In practice, most doctors prefer to 'overtreat' by administering a dose that causes the thyroid to become completely inactive – what doctors call an ablative dose – on the grounds that an underactive thyroid is easier to treat than an overactive one, although this, as we've seen, is debatable.

You will be invited for a check-up one to two months after treatment. You will be put back onto antithyroid drugs, but these will be phased out once your thyroid function is restored to normal. Thyroid function will be tested again four to six weeks after this. If your thyroid is found to be still overactive, you may need a further radioiodine treatment. Around one in 10 of those treated need a second treatment within a year, although only a very small percentage require a third treatment.

You will need regular check-ups – preferably at least once a year – for the rest of your life to keep track of how your thyroid is behaving.

## Is It Safe?

The million-dollar question is, of course, whether there are any risks attached to radioiodine therapy. Many people – understandably, given the known links between exposure to radiation and cancer – are under the misapprehension that $^{131}$I therapy can cause cancer or leukaemia. This erroneous fear should be laid to rest. Radioiodine treatment has been used for over 50 years with no evidence of an increased risk of cancer or leukaemia.

However, the most contentious issue among the endocrine community is whether radioiodine treatment can cause thyroid eye disease to worsen (*see Chapter 8*). Occasionally, patients treated with radioactive $^{131}$I develop harmless nodules a few years after treatment, and sometimes the treatment causes an inflammation of the thyroid (thyroiditis) which, fortunately, is usually temporary.

Even without any danger of serious consequences, however, there's no doubt that radioiodine treatment can be stressful, as Maggie, 34, discovered:

*I found the treatment especially traumatic. I developed diarrhoea and sickness, nose bleeds and sore gums, and my skin looked grey and ashen. But it was the emotions that surfaced that caused me most pain. I was put in isolation, away from the other patients, and I went through every feeling imaginable – depression, anger, the lot. It was horrible not being able to have any contact with anybody. I knew the treatment was necessary, but I had problems with the idea of it. I used to be an active antinuclear campaigner and now here I was drinking the stuff.*

*The second time I had to have radioactive iodine, I spent two weeks in isolation at home. It was totally different. Even though I couldn't sleep in the same room as my partner, I was with my family, I could make phone calls, walk around the garden and feel much more normal.*

## Watchpoints

- Although radioactive iodine is safe if your thyroid is overactive, it can affect the thyroid function of those with normal thyroids. For this reason, you will need at least three days off work, depending on your job.
- You should avoid non-essential close contact with babies, young children and anyone who is pregnant for up to four weeks.
- If you are being treated with lower doses, you may be treated at home, provided you undertake to limit your

contact with others. This means minimizing the amount of time spent in public places and travelling by public transport (usually for up to four days), keeping your distance from babies or young children, and avoiding kissing and other close contact with your partner, friends or relatives – especially women of childbearing age – until the radioactivity has cleared from your system.

- Smaller doses can be taken over two to four weeks and without the precautions mentioned above, so long as your job doesn't involve contact with radiation. Even then, try to avoid carrying babies or young children for a few days after treatment.

- Radioiodine treatment is stressful. Give yourself little treats like a new book to read, a foot or shoulder massage with aromatherapy oils, or a good video to watch to help you cope.

- Plan some bigger treats that you can do with others when you are clear of the radioactivity – a weekend away with your partner or a friend, a short holiday, or a professional massage, pedicure or facial.

Note: The duration of time that precautions will apply depends on the amount of radioiodine the doctor has prescribed following a scan, and should be explained to you before your treatment. If they aren't, you need to ask.

## Treatment for Thyroiditis

In mild cases of viral thyroiditis, no treatment may be necessary or, alternatively, anti-inflammatory drugs like aspirin can help. If you are in prolonged pain, the doctor may prescribe steroids for a month or so, gradually tapering off the dose as you get better.

If your thyroid has been damaged as a result of autoimmune thyroiditis, you will need to be treated for hypothyroidism (*see page 83*).

**Postpartum Thyroiditis (PPT)**

*See Chapter 9.*

## Goitres, Nodules and Cancer

If you have a goitre, particularly one that is small, soft and causing no symptoms, no treatment is usually necessary. However, if you have developed a single 'hot' nodule or a multinodular goitre, you may be offered a partial thyroidectomy, or lobectomy, to remove the overactive part of the gland. Alternatively, you may be offered radioactive iodine therapy. Surgery may also be performed if you have a thyroid adenoma.

Treatment for thyroid cancer involves surgery and/or radioactive iodine. Once the thyroid is removed or destroyed by the radioactive iodine, you will become hypothyroid and need to take thyroxine ($T_4$). This will usually be given at a dose high enough to completely suppress the production of thyroid-stimulating hormone (TSH) by the pituitary gland so that it is undetectable on thyroid function tests. You will need regular monitoring after treatment to make sure the cancer doesn't return or spread.

Ideally, you will be referred to a centre with a multidisciplinary team with expertise in the diagnosis and treatment of thyroid cancer. Such a team would include a surgeon, endocrinologist and oncologist (cancer specialist), and have the support of a pathologist (specialist in cell analysis), a medical physicist, biochemist, radiologist and specialist nurse. Unfortunately, this tends not to happen in most places in the UK, and it is likely to be several years before it does. At the very least, there should be good communication between the various professionals involved in your treatment. Before having treatment, you should be given full details of what to expect from treatment. If you aren't, then don't be afraid to ask.

# Thyroid Eye Disease

This condition is covered in detail in Chapter 8.

# Treatment Summary

## Table 5.2
## Summary of treatments for thyroid problems

| Treatment | What It Does | When may it be used |
|---|---|---|
| Antithyroid drugs (thionamides) | Quell excess production of hormones | Hyperthyroidism, Graves' disease, postpartum thyroiditis (PPT) |
| Beta-blockers | Lower heart rate | Hyperthyroidism, PPT |
| Thyroid replacement therapy ($T_4$, $T_3$, $T_4/T_3$, animal thyroid extract) | Replaces one or more thyroid hormones | Hypothyroidism, Hashimoto's disease, PPT, after radioiodine therapy or surgery (for cancer or hyperthyroidism) if hypothyroidism occurs, after autoimmune thyroiditis |
| Radioiodine therapy | Kills thyroid cells | Hyperthyroidism, Graves' disease, after thyroid surgery, single hot nodule, thyroid cancer |
| Thyroidectomy (surgery to remove all or part of thyroid) | Removes some or all overactive thyroid cells, thus lowering production of thyroid hormone | Hyperthyroidism, Graves' disease, thyroid nodules, multinodular goitre, thyroid adenoma, thyroid cancer |
| Corrective eye surgery | Improves physical symptoms of thyroid eye disease (by enlarging orbit or correcting lid lag) | Thyroid eye disease (see also Chapter 7) |

## A Raging Calm?

A great deal of controversy surrounds the diagnosis and treatment of so-called mild or subclinical thyroid disease. Over the past few years, a host of influential popular books has appeared, attributing symptoms such as low energy and fatigue to mild hypothyroidism. According to their authors, including medical journalists, endocrinologists and complementary health practitioners, these 'silent' thyroid illnesses can underlie or exacerbate a huge number of problems of the 21st century as they can be masked by or masquerade as other complaints such as obesity, anxiety and depression, arthritis, digestive problems, high cholesterol, low sex drive, infertility and menopausal problems and – perhaps most controversial of all – chronic fatigue syndrome and fibromyalgia (debilitating muscle and joint pain accompanied by fatigue and depression).

Orthodox doctors and endocrinologists normally define subclinical hypothyroidism as having normal or borderline levels of thyroid hormone, but with raised levels of thyroid-stimulating hormone (TSH), suggesting that the thyroid is beginning to fail. These conventional doctors acknowledge that subclinical hypothyroidism exists. Where many of them part company is over the issue of whether and how they think it should be treated.

Many doctors in the orthodox camp remain cautious when interpreting slight discrepancies in hormone levels and treating subclinical problems, especially when the results of thyroid function tests show up as 'normal' or 'borderline' and symptoms are few.

By contrast, advocates of the other school of thought argue that the current tests are a blunt instrument for detecting thyroid problems. They believe that people with suspicious symptoms should resist what Richard L. and Karilee Halo Shames, authors of *Thyroid Power* (HarperCollins), describe as the 'tyranny of the tests' which, they believe, leaves many thousands of people with low thyroid function untreated. According to the Shames and other physicians, treating the symptoms of mild hypothyroidism can often bring about dramatic improvements in health and wellbeing. Even more contentious is the recommendation to use high doses of thyroxine to treat mental health problems, such as depression and the extreme mood swings of manic-depression.

Other experts counter that giving thyroxine to those whose test results fall within the 'normal' range can, at best, offer little or no benefit and, at worst, may be unsafe. There are concerns that it may disrupt the heart rhythm as well as fears of long-term effects on the brain, leading to a need for thyroxine indefinitely – in effect, becoming addicted to it. Dr Charles Shepherd, a British endocrinologist, has expressed particular concern as he believes that many of the benefits experienced by patients taking thyroxine may be a result of its stimulating effect rather than because they are suffering from low thyroid function.

A similar debate is ongoing as to when or whether early subclinical or mild hyperthyroidism should be treated. As with subclinical hypothyroidism, a proportion of cases of mild overactivity may progress to full-blown hyperthyroidism.

The general consensus, outlined by Dr Anthony Toft in the *New England Journal of Medicine* in August 2001, seems to be that, provided the condition is straightforward – no nodular thyroid disease or measurable levels of excess thyroid hormone – treatment is not necessary. Dr Toft also recommends having thyroid function tests every six months to keep an eye on the condition's progress. In those with what Toft describes as 'questionable' symptoms – such as fatigue – he suggests a six-month trial of an antithyroid drug at a low dose and, if this is successful, then treatment with radioiodine. Where subclinical hyperthyroidism is due to nodular thyroid disease, he says treatment is 'more routinely justified' because, in this case, the condition often progresses to overt hyperthyroidism.

So who is right, and how can you decide? The truth is, it's difficult for all of us, especially those of us who don't have a background in science, to make decisions and choices about our health, especially when the medical approach appears to

change from day to day. The practice of any science – and medicine is no exception – is constantly changing so that, often,
yesterday's controversial theory is tomorrow's orthodoxy.
Other theories will be dropped by the wayside as more scien-
tific knowledge accrues, and fashions in treatment, as with any
other fashions, will come and go.

But this doesn't mean that you shouldn't be an active
participant in your treatment. You deserve to know why a
particular treatment is being suggested or withheld, and it is
important to keep an open mind. If treatment is taking a long
time to take effect or isn't having any effect, it can be tempting
to clutch at any straw in the wind.

However, the history of medicine is littered with disasters
caused by the adoption of therapies that weren't sufficiently
tried and tested, and many of these have affected women.
Unless there is strong evidence of a treatment's efficacy and
safety, it doesn't hurt to be sceptical. Above all, it's up to all of
us to keep informed and learn as much as we can about the factors affecting our health, and to find doctors whom we can
trust, and who are prepared to listen and work alongside us to
devise a management plan that is both safe and effective.

As you have seen, treating thyroid problems is not always
easy. The important thing is not to despair: with patience and
persistence, you can get your thyroid under control and live a
fuller, healthier life.

# CHAPTER SIX

# I Just Want to Feel Normal Again

As far as many doctors are concerned, thyroid disease is fairly straightforward – you do some tests, reach a diagnosis and treat the problem – end of story. But from the other side of the doctor's desk, the outlook is often very different. Although it may be a tremendous relief to find out what's wrong, a diagnosis of thyroid disease is usually just the beginning. Learning to live with what is a long-term and frequently variable condition can be an immense challenge.

Apart from anything else, discovering that you have a chronic illness is often a blow to your self-confidence and self-esteem. You may come to distrust your body and feel alienated from it for behaving so unpredictably. The realization that you are likely to need medication and regular check-ups for the rest of your life can make you feel trapped. You may resent your dependence on the medical profession and others in general, and mourn your loss of freedom. Although those caring for you may be trying to be kind, they may simply be unaware of how you are feeling and the distress you have bottled up inside.

Maggie, 34, who has an overactive thyroid, speaks for many women when she says:

> I can't believe it has come from nowhere, and it's taken over my life. I'm taking pills every day.
> I feel totally out of control. I never imagined it

*would be like this. I just want to be well again.*
*I'm sick of pills and hospital.*

## Living With Loss

Every time we go through any major life change, we experience loss – and with this comes grief. In the case of thyroid disease, there are many losses to deal with. As well as the physical suffering, there is the loss of your self-image as a fit, healthy person. The physical effects of thyroid disease can be especially hard to bear, given the appearance-conscious society we live in. If you have severe eye problems, or have put on a lot of weight or are losing your hair, you may feel deeply unattractive. In a culture where being overweight is deemed not just unhealthy, but unappealing, even a few extra pounds can seriously dent your body image. As Angela describes:

> *I am painfully aware of society's disapproval of*
> *my abundant flesh. I find myself rushing at the*
> *earliest opportunity to tell acquaintances that I'm*
> *fat because I'm ill, not because I overeat, honest.*
> *Some people are sympathetic, but a woman once*
> *said I was using my illness 'as an excuse'. I don't*
> *eat much at parties in case someone is watching*
> *me.*

Although losing weight is generally viewed as desirable, being 'too thin' can also draw censure – witness the regular articles in the popular press about whether an actress or singer who has lost a lot of weight has anorexia. One doctor, writing in the *British Medical Journal*, described the case of Anne, a hypothyroid running champion whose descent into Addison's disease [an autoimmune disorder affecting the adrenal glands that, though rare, is known to affect some people with thyroid problems] went unrecognized by friends and medical staff alike until she was at death's door:

*People said she was a nutcase, that she had*
*invented the running glories and nicked the*
*trophies. Perhaps she had anorexia nervosa.*
*Perhaps, I wondered, her thyrotoxicosis was*
*fictitious. The more we labelled her problem as*
*functional [having no physical cause], the more*
*she fitted the mould. I still have the article clipped*
*from a psychiatric journal that seemed to describe*
*her symptoms and behaviour perfectly.*
  *It's a cautionary tale that demonstrates just*
*how all of us, including members of the medical*
*profession, can be affected by preconceptions.*

If you have a large, noticeable goitre or if you have to have surgery, these too can result in loss of self-confidence, and feelings of loss and grief as can medical problems associated with thyroid disease, such as difficulty conceiving or miscarriage. If children had always been part of your life plan, you may feel 'less of a woman' and a sense of loss of purpose in life. You may feel as though it's your 'fault' if you and your partner can't have children. Likewise, if your baby has a birth defect or a problem with his/her own thyroid, this too can lead to feelings of guilt.

Thyroid disease in itself can cause you to be moody and, at the same time, lead to a loss of mental sharpness and lack of interest in life. Rapid fluctuations in thyroid levels are particularly liable to lead to mood swings. Also, some of the treatments used for thyroid problems can have psychological effects; for example, beta-blockers can make you feel tired and slow. Even when you have been successfully treated, it can take some time before you feel yourself again and experience a real sense of health and wellbeing.

## All Nicely Under Control?

Unfortunately, women with thyroid disease often see things in a different light from their doctor's, as evidenced by Judith's comments:

> *Most of the time I work from home, which is fine, but every so often, I have to go to town for a meeting. I get terribly panicky about it, wondering whether I'm going to be okay that day. In one way, being self-employed helps – I can take it easier when I have to – but, on the other hand, if I don't work, I don't get paid and that's a real worry. I suspect if I had a regular job, I'd have had to take a lot of time off. My consultant said to me, somewhat casually, 'It'll probably end in surgery'. All the time, I was thinking, 'How am I going to fit that in?' As soon as they say, 'It's very common among middle-aged women', it's as though it's not serious. The other week, I saw a doctor for a check-up who said, 'Oh, it's all nicely under control, isn't it?' I said, 'No, it isn't'.*

## Seeking Support

Doctors who are experienced in treating thyroid problems ought to be aware of such factors, but it's a sad fact that not all members of the medical profession are as sympathetic as they could be. Many women feel that their complaints are not taken seriously, so you may find that talking about your problems with others who have thyroid disease – either face-to-face, over the phone or through one of the Internet chat rooms for people with thyroid problems – is helpful. If you are truly depressed, then counselling, psychotherapy or antidepressant medication may tide you over a rocky patch and help you regain your emotional equilibrium. In other cases, depression can be lifted by changes in medication to achieve better control over symptoms.

## Treatment Trials

Medical diagnosis and treatment can also take its toll. Hospitals are institutions – large, complicated and hierarchical. In a busy clinic or ward bustling with activity, it's easy to feel lost, alone and isolated. Away from your own familiar surroundings and in the grip of an illness that is hard to understand, you may feel terrifyingly powerless and depressed. Finding out more about your illness through reading may enable you to discuss your treatment in a more informed way with your doctor and help you to feel more in control. However, some patients prefer to leave medical decisions to their doctors. If this applies to you, accept it and do what feels right for you.

## Not In the Textbook

Time and again, women suffering from thyroid problems complain that they felt their needs and wishes were ignored or trivialized because they weren't 'textbook' cases. As recalled by Maggie:

> After I took the first tablet, I took my baby to the park. I was sitting in a cafe having a coffee when, suddenly, I experienced a terrible hot flush and a prickling sensation in my body. I felt really peculiar. The sensations came in waves. Then, when I went to bed, I had muscle spasms in my thighs. I felt really panicky. Another time when I was in the post office, I came over all faint. I was very frightened, wondering how on earth I was going to get home. Eventually, I went back to the

*doctor, who told me it was psychosomatic. He said*
*that it took three months to get a response to*
*thyroxine and that I couldn't possibly be having a*
*response in three hours. I was in floods of tears.*
*I could see that he just thought I was being*
*neurotic.*

   *Eventually I got the doctor to agree to give me*
*a blood test before the usual three months [the*
*time the dose normally takes to build up to its*
*optimum level]. My thyroid count had gone up*
*almost to normal. They were astonished at how*
*quickly my blood had taken it up, and I felt that*
*vindicated me.*

From the medical staff's point of the view, so long as your
hormone levels are within the expected range, there may be no
problem. But as Judith observes, an illness that can lead you to
feel perfectly well one day and terrible the next can be particularly undermining because you never know what to expect. For
women who are used to managing their own lives, everything can
appear frighteningly out of control, even when, medically, the
doctor perceives treatment to be proceeding well.

## Personality Counts

Feelings can be a particular problem with any thyroid condition
as the illness itself can lead to changes in behaviour and personality. Time and again, women with thyroid problems describe
behaviour and feelings that are uncharacteristic – becoming
lethargic and depressed when they are normally energetic and
lively, arguing with people close to them when they are usually
placid, being unable to think straight when they are normally
mentally sharp. All of these are part of your condition, so don't
blame yourself too much if you experience them. Being aware
of your emotions can help you to control your behaviour by,
for example, employing anger-management techniques such as

walking away from an inflammatory situation or taking deep breaths.

## Regaining Confidence

It may be a while before you learn to trust your body again. Accept that it takes time, but believe that, as you learn to manage your condition, you will be in a better position to deal with all the changes it has brought into your life. You may find that as you learn to manage your condition, you also learn new ways of thinking about it. If you can see your problem, however unwelcome, as opening up the prospect of positive growth and development, which may make it easier to live with. Many self-help and voluntary organizations have been started by people who had feelings they needed to deal with and experiences they felt would benefit others. Lyn Mynott, who runs the self-help group Thyroid UK, is one of them:

> *Although I would not have chosen to be ill and have all the years of pain, if I had not been ill, I would not have decided to take on a job that I could do at home – teaching English to students of other languages. I loved it so much that I took more qualifications and now teach in an adult-education college, teaching all kinds of people from other countries – refugees, au pairs, business people.*
>
> *Also, if I had not been ill, I would not have set up Thyroid UK and be helping the hundreds of people out there who need information. I like to think that, through my illness, many people's lives have changed for the better.*

Psychologists describe several stages in learning to deal with life transitions like the development of a chronic illness. At first, there is immobilization – you feel numb, out of control and unable to act. Next comes minimization – you describe your problem as trivial or deny it altogether. This is then followed by a period of self-doubt and depression as the reality sinks in. The low point in self-esteem is acceptance or letting go, when the reality of your new situation really hits home and you begin to realize the limits it may place on the way you lead your life.

As with many crises, once you hit rock-bottom, you begin to go up again. Self-esteem begins to rise as you test your new situation and start to make positive changes in your life. Such changes can be something simple, like joining an exercise class or deciding to eat more healthily. Angela writes:

> *Being ill has revealed to me a resilience I never knew I had, has taught me to value the good things in my life – my children, my friends. Now I take nothing for granted, no one at face value. I have gained a tiny insight into the even more serious illnesses that befall people and recognize their real courage in adversity. I begin to understand the isolation that can be felt by the differently abled in a torso-obsessed culture.*

Such insights help you to 'internalize' the changes, make them part of you and move on, a changed but enriched person.

## Dealing With Negative Emotions

Many psychologists believe that the way you think plays a large part in determining how you feel. Negative thoughts can drag you down, and make you feel tired and unenthusiastic whereas making an effort to think more positively can lift mood and make you feel more energetic. Some people find that replacing negative, undermining thoughts with positive ones can make a real difference in the way they deal with their thyroid problem. Keeping a thoughts diary to identify unhelpful

patterns of thinking may be useful. If you find yourself thinking 'this is awful; I'll never get any better', replacing it with the
more realistic 'some days are bad days, but I am having treatment and doing things to help myself feel better' can make a
big difference in how you feel. Others claim that using positive
'affirmations' such as 'I am in control of my thyroid', 'I won't
let this beat me' or 'I am still me' and repeating it often to
themselves is helpful.

## Dealing with Practicalities

Taking action is another way to defuse stress and help yourself
feel more positive. Discovering as much as you can about your
condition, and how it is diagnosed and treated will enable you
to understand your doctor's point of view and allow you to be
more fully involved in its management. Keep a written record
of physical symptoms and note any medical treatments you are
taking, together with dates and results of any investigations
and other procedures you have undergone. If you are unhappy
with your treatment, write a letter of complaint. If you still
don't feel well despite treatment, find out more about the
different therapeutic options available and make an appointment to see the doctor to discuss these. If you don't receive any
joy from your doctor, consider seeking a second opinion.

Christine, who has had an underactive thyroid for 29 years,
says,

> *The important thing is not to give up hope
> because, sometimes, it takes a long time to feel
> better. My advice to others in my position is don't
> expect to feel well in a couple of weeks. It can take
> months or even years before you get your strength
> back. I have to monitor my energy levels because,
> if I overdo it, my batteries run out.*

## Keeping an Eye on Your Thyroid

Dr Rowan Hillson, in her book *Thyroid Disorders*, suggests measuring your neck monthly if you have a goitre (see below) as well as keeping a check on your weight and taking your pulse every so often to confirm its regularity or whether it's getting faster or slower. She also counsels becoming aware of changes in your physical appearance such as to your hair, skin, nails and eyes, together with other tell-tale signs of changes in thyroid activity, such as feeling unusually hot or cold, changes in your appetite, bowel motions, menstrual patterns, any unusual fatigue or excessive energy, muscle weakness or fatigue, depression, anxiety, or changes in mental function such as forgetfulness or a lack of concentration. She writes:

> It can be difficult to recall what happened, when and which pills you took for how long. But if you have written notes of things you can measure, like your weight and pulse, as well as how you feel, you can see if your thyroid function is speeding up, slowing down or steady.

Such activities, as well as being useful, can help you feel much more in control. It's easy to go on to autopilot when receiving medical treatment. By monitoring your condition and participating in your treatment, you will feel much more in charge.

## The Thyroid Neck Check

As part of its 'Think Thyroid' campaign, the American Association of Clinical Endocrinologists suggests that anyone with symptoms of thyroid activity (see above) or a family history that puts you at risk should do a simple neck check. All you need is a glass of water and a hand-held mirror.

- **Step 1.** Hold the mirror and focus it on your neck, just below the Adam's apple and just above the collarbone, where your thyroid is situated.
- **Step 2.** Focusing on this area, tip your head back.
- **Step 3.** Take a swig of water and swallow it.
- **Step 4.** As you swallow, observe your neck. Check for bulges or protrusions as you swallow, being careful not to confuse your Adam's apple with your thyroid (the thyroid is further down the neck, closer to the collarbone). You may want to repeat this several times to be sure.
- **Step 5.** If you notice any bulges or protrusions, you may have a thyroid problem that needs investigating. Make an appointment to see your GP.

Adapted from the AACE's *How to Take the Thyroid 'Neck Check*™, and reprinted, with permission, from the American Association of Clinical Endocrinologists.

## Check Up on Your Genes

Knowing your family medical history can help you determine if you or other members of your family are at risk of thyroid problems.

### DRAW UP A FAMILY TREE
Ideally this should cover three generations on both sides of your family. Jot down the age and onset of all the conditions that your relatives developed.

### LOOK FOR PATTERNS
These include an early onset (before age 50) and multiple family members being afflicted with the same disease or a particular type of diseases. It's also worth noting other conditions such as osteoporosis and heart disease, which can be associated with thyroid conditions.

Other clues to consider include whether or not you or other members of your family have/had:

- an underactive or overactive thyroid
- premature (in your 20s) grey hair, a sign of autoimmunity
- a history of other autoimmune disorders such as type 1 (insulin-dependent) diabetes, rheumatoid arthritis, pernicious anaemia or vitiligo
- a collagen disorder affecting the connective tissues and causing symptoms such as thin, slack skin and poor wound-healing
- polyglandular deficiency syndrome (when several glands are underactive)
- thyroid cancer.

If you detect a pattern, other members of your family may be at risk. Be aware of potential symptoms so that you can get an early diagnosis, and make sure that other members of the family know they are at risk and what they should watch out for. If there is thyroid cancer in your family, make an appointment to see the doctor who can, if necessary, refer you for genetic testing and possibly prophylactic thyroidectomy.

## Eating For Your Thyroid

When you are feeling depressed or tired, it may be hard to summon the energy to make healthy food choices and prepare nutritious meals. You may get into the habit of buying ready-made meals or reaching into the freezer cabinet rather than planning and cooking food from fresh. Although convenient, this can be counterproductive as poor food choices can cause you to pile on the pounds and lead to a greater lack of energy. In addition, your condition or the medications you are taking may alter your appetite and need for certain nutrients.

Making the effort to eat a healthy diet may not only help you feel better physically, it can also enable you to feel more in control of your life. Many people imagine that eating a healthy diet means having to buy special foods or supplements or making big changes in what they eat. In fact, taking a few small, achievable steps can do a lot to increase your zest and sense of wellbeing.

- **Stay balanced.** The best way to ensure that you get a good range of nutrients is to eat a balanced diet. That means being aware of what you eat. If you do occasionally succumb to foods high in salt, fat or sugar, balance them out with healthier choices.
- **Watch your portions.** Portion size is one of the keys to weight control. Eating reasonable portions will enable you to include some of the foods you enjoy and still eat healthily. Portion control doesn't mean counting calories or having to weigh everything you eat. One useful rule of thumb is that no portion on your plate should be bigger than the palm of your hand.
- **Variety performance.** Eating a wide variety of foods is another way to ensure you get a full complement of nutrients. Many of us get stuck in a rut and eat the same old foods day in and day out. In fact, one recent survey showed that, in any given year, what we put in our shopping baskets varies by less than 10 per cent. With the vast range of ingredients available in the shops today, there is more choice than ever before. Try to increase the variety of your diet by trying out new foods and ingredients.
- **Eat breakfast.** Breakfast is the most important meal of the day both to provide you with nutrients after having spent eight hours or more without food, and energy to fuel you through the day. Eating a sustaining breakfast can also help with weight control. Good choices include porridge made with water and sprinkled with chopped nuts, low-sugar muesli, fruit with low-fat yoghurt, a fruit smoothie or wholemeal toast with a slick of yeast extract or fruit spread.
- **Eat little and often.** Eating small frequent meals is better for energy levels – and your weight – than eating three big, heavy meals. Snacks can help keep you satisfied between meals, but make sure they are healthy. Choose

fresh fruit, vegetables such as carrot and celery sticks, nuts and seeds or small amounts of dried fruits rather than cakes, sweets and biscuits.

- **Eat regularly.** Try not to skip meals as this can lead to low blood sugar, which aggravates tiredness and lack of energy. Foods like grilled or baked chicken, or fish brushed with oil, lemon and a few herbs, can be prepared in minutes. If you are vegetarian, make use of the vast array of pulses now available. Canned ones are convenient, but investing in a pressure cooker means dried pulses can be cooked in minutes. These combined with a packet of salad, or a few grilled or steamed vegetables, produces a virtually instant meal.

- **Cut down on caffeine.** Because they are stimulants, it can become easy to rely on coffee, tea and other caffeine-containing drinks to give a quick boost to your low energy. Unfortunately, too much caffeine is overstimulating and makes you feel nervous and edgy as well as making it difficult to sleep at night. Caffeine can also exacerbate symptoms of premenstrual syndrome (PMS), which can be a particular problem for women with thyroid problems, and there is also some debate over its effects on fertility. There is also research to suggest that inflammatory reactions are more common in heavy coffee-drinkers – all good reasons to cut it out or cut down. Try to stick to one or two cups of coffee or tea a day and don't drink caffeinated beverages after lunchtime (to avoid it keeping you awake at night). Substitutes include mineral water, and herb and other teas such as rooibosch (a South African tea that has a tea-like flavour, but no caffeine).

- **Moderate your alcohol intake.** Alcohol contains empty calories that you don't need if you're watching your weight. Alcohol can also sap energy and interfere with sleep. Try to keep your alcohol intake to under 14 units a week (one unit = a glass of wine, a single measure of spirits or a half-pint of beer) or, preferably, far fewer.

- **Drink two litres of water a day.** Water is needed to hydrate your body and give you energy, and also helps weight loss. You need to drink more if you take a lot of exercise.

## Balancing Your Diet

So what foods should you be eating, and how can you ensure that your diet is healthy? To work at its peak, your body needs a wide range of different nutrients. A healthy diet should provide your body with these nutrients to produce energy for fuel to ensure that your body meets the demands of daily life and keeps your immune system functioning optimally.

### CARBOHYDRATES

Carbohydrates (starchy foods) are burned easily by the body and are particularly vital for energy. However, not all carbohydrates are created equal. Simple carbohydrates – found in sugary, processed foods like sweets, ice cream and biscuits – cause a rapid rise in the hormone insulin, produced by the pancreas, providing your body with a quick rush of energy. Unfortunately, this can lead to energy dips and tiredness as the sugar is quickly used up.

Complex carbohydrates – found in wholemeal bread, wholegrains, fruit and vegetables, and pulses, wild rice, millet and wholemeal pasta – result in a more gradual release of insulin, providing more sustained levels of energy.

If you have a thyroid problem, it is especially important to choose complex carbohydrates not only because insulin is involved in appetite control, but because there is evidence that high levels of insulin block the body from burning up fat for energy, thus making it more difficult to rid yourself of excess pounds.

Fibre is another type of carbohydrate. It comes in two forms: soluble fibre, found in pulses, and many fruit and vegetables; and insoluble fibre, such as bran in the husks of grains, and the fibrous parts of fruit and vegetables. Soluble fibre is now considered to be the most important type of fibre as it is helpful in lowering cholesterol levels, an important factor for anyone with a thyroid disorder due to its links with heart disease.

## FATS

Just as there are two kinds of carbohydrates, so there are also two kinds of fats: the 'bad' saturated variety, found in animal products and dairy foods; and the 'good' unsaturated fats, found in olive oil, nuts, seeds and oily fish. These are a source of essential fatty acids (EFAs), which are vital for the healthy function of cells.

Controlling your fat intake is essential both for weight control and to help reduce heart disease. However, it is important not to cut out all fats. Bad fats slow your metabolism and encourage harmful LDL (low-density lipoprotein) cholesterol to accumulate on the walls of the arteries, increasing your risk of heart disease and stroke. Good fats, by contrast, are vital for energy and your immune system. They help speed up your metabolism and also sweep away bad LDL cholesterol, so reducing your risk of heart disease. And last but not least, they can improve the condition of your skin, hair and nails – important benefits for people with thyroid problems.

## VITAMINS, MINERALS AND TRACE ELEMENTS

Vitamins and minerals are nutrients that are found in small quantities in food and are vital for the proper functioning of the body. Of particular importance to health and wellbeing are the so-called antioxidant vitamins and minerals, sometimes known as the ACE nutrients – vitamins C, E and beta-carotene, which is converted to vitamin A in the body together with the trace element selenium – found to help improve the health of the immune system and protect against a number of degenerative illnesses and conditions, including cancer, heart disease, diabetes, arthritis and ageing, all of which are relevant to the thyroid. Recent research suggests that a shortage of selenium can contribute to thyroid problems.

## Special Nutritional Issues

According to nutritional doctor Stephen Davies, people with an overactive thyroid have an increased demand for certain nutrients, especially the B-complex vitamins, which combat stress, and trace minerals. There is sometimes an association between B-complex vitamins and fatigue. Also, vitamin $B_{12}$, found in liver, beef, pork, eggs, milk, cheese and kidneys, is particularly important in combating pernicious anaemia, which those with autoimmune thyroid disease are at a higher risk of developing. $B_{12}$ is also involved in focusing and visual sensitivity, which may be relevant to sufferers with thyroid eye disease.

Calcium, of course, is particularly important to women with thyroid problems because of the increased risk of osteoporosis (*see Chapter 10*). Calcium is also important for a regular heartbeat, metabolizing iron and aiding the transmission of nerve impulses. As we have seen, calcitonin – which is involved in regulating calcium levels in the blood – is one of the thyroid hormones. Magnesium, found in figs, lemons, grapefruit, sweetcorn, almonds, nuts, seeds, dark-green vegetables and apples, is also necessary to help metabolize calcium. Magnesium is also an aid in fighting depression and helping to promote a healthier cardiovascular system.

Manganese – found in nuts, green leafy vegetables, peas, beetroot, egg yolks and wholegrain cereals – is another vital mineral involved in the formation of thyroxine. It can help eliminate fatigue, aid muscle reflexes, improve memory and reduce irritability. You can get all these nutrients by paying close attention to your diet. In addition, you may want to consider taking a daily multivitamin/mineral and/or a vitamin B-complex supplement.

## The Glycaemic Index

The glycaemic index (GI; *see Table 6.1*) rates the food we eat according to the speed at which blood sugar and insulin levels rise. High GI foods – such as potatoes, bananas, white bread, cereals and honey – lead to a sharp rise in energy as a result of the release of insulin. An increasing number of nutritionists and fitness experts believe that a diet with too many high GI foods contributes to tiredness and lack of energy, and can make it difficult to lose weight – a theory which, if true, has obvious implications for persons with thyroid problems. In contrast, medium-to-low GI foods, such as apples, not only maintain energy levels for longer and make it easier to lose weight, but also increase the production of the brain chemical serotonin, which helps control appetite and improves mood. Try to follow a diet mainly made up of low-to-medium GI foods and keep the high GI foods for a special treat.

---

### Table 6.1
### Foods according to their glycaemic index (GI)

| High GI foods | Medium GI foods | Low GI foods |
|---|---|---|
| Bananas | Beetroot | Apples |
| Biscuits | Bran-based cereals | Apricots (fresh) |
| Bread | Brown rice | Beans & pulses |
| Cooked carrots | Grapes | Green vegetables |
| Cornflakes | New potatoes | Peaches |
| Honey | Oats | Peppers |
| Parsnips | Pineapple | Sweet potatoes |
| Popcorn | Strawberries | Tomatoes |
| Syrups | White pasta | Wholegrain rye bread |
| Sweetcorn | Yams | |
| Watermelon | | |
| White bread | | |
| White old potatoes | | |

## The Acid–Alkali Balance

Another important nutritional issue, given the potentially increased risk of osteoporosis in women with thyroid disease, is what nutritionists call the acid–alkali balance of the body. A number of studies suggest that eating foods that create acidity in the body is associated with lower bone density. By contrast, alkali-forming foods help to conserve bone. The typical Western diet is extremely 'acid'. What's more, we become more 'acidic' as we age while our kidneys become less effective in ridding the body of excess acid. Acid-forming foods include butter, cheese, cream, full-fat milk, white bread and pasta, basmati rice, pastries, cakes and biscuits, red meat, processed foods, chocolate, tea, coffee and alcohol. Acidic fruits include citrus fruits, cranberries and tomatoes, and nuts such as cashews, pistachios, peanuts and pecans are also acid-forming. Most other fruits and vegetables are alkali-forming.

## Foods to Eat With Caution

As we've already seen, iodine is essential for the production of thyroid hormone. Some foods – known as goitrogens – have the capacity to obstruct the body's uptake of iodine. Consuming too many of these foods can make an underactive thyroid worse. They include brussels sprouts, broccoli, cabbage, millet, peanuts, pine nuts and turnips. If you've been diagnosed with hypothyroidism, it may be advisable to limit your intake of these foods.

# A Sense of Balance

It's especially important if you are ill to pace yourself and balance activity with rest and relaxation. This doesn't mean that you should rest all the time. Exercise can be an important part of the solution for many women with thyroid disease who complain of tiredness and lack of energy. Exercise increases the capacity of your heart and lungs, and can help you regain the energy to do more.

Another common complaint is difficulty in losing weight. The answer to this is to make sure you are on the correct dose of medication and to combine sensible eating with an exercise programme. Such a programme should include fat-burning aerobic exercise such as running, walking, swimming or anything that pushes your heart rate up and resistance work using weights or your body's own weight in exercise systems such as yoga or Pilates. Muscle is more metabolically active than fat, which means that if you make resistance exercise a part of your regular programme of activity, your body will burn more fat – even when you're just sitting! And even if you don't lose weight with exercise, your muscles will be more toned so that your clothes will fit better.

Depression is another common consequence of thyroid problems and, again, exercise can help. Sustained activity (around 45 minutes or more) triggers the release of endorphins, hormone-like brain chemicals which help you feel more cheerful, calmer and more relaxed.

Above all, exercise can help prevent some of the more serious health consequences of thyroid disease such as heart disease and osteoporosis. Of course, if it has been a while since you've taken exercise, you're unlikely to be able to exercise vigorously straightaway. However, if you start slowly and build up gradually, you will certainly be surprised at how much you will eventually be able to achieve.

It's not always easy to begin exercising as your muscles may be weak as a result of your thyroid disorder, and you may tire easily. But avoiding exercise is not the answer as it can lead to the body becoming 'deconditioned', with the result that you have no spare energy reserves and even everyday activities become a major effort. Then, when you do try to exercise, your joints may ache, and you may become easily tired and breathless.

This vicious, downward spiral may be reversed by becoming more active gradually. Gentle activities, such as walking, swimming or yoga, can be especially invigorating if you are feeling tired. However, any such exercise needs to be tailored to you as an individual, so take expert advice from your doctor, physiotherapist and/or a fitness specialist before embarking on any exercise programme.

## Exercise Tips

- **Pick an activity you enjoy.** You're more likely to carry on exercising if you are doing something you like. Exercise doesn't have to mean joining a gym, although many people find this is a great way to get started as it provides a dedicated space and time in which to exercise, and the support of professionals who can help devise an effective exercise programme. If you don't like the idea of a gym, there are plenty of other options to choose from. If you are short of ideas, think back to what you used to enjoy at school – perhaps you were good at tennis or enjoyed swimming. If so, then think about joining a tennis club or making a date at your local swimming pool. If you've never been a sporty type, then consider joining a dance class. There's a vast number on offer, including ballet, contemporary, flamenco, salsa, samba, tango, rock n' roll, line dancing and street dance, and all of them offer a terrific aerobic workout to help maintain a healthy heart and burn fat.
- **Find an exercise partner.** Exercising with a friend can aid motivation and give you less opportunity to opt out when you just don't feel like it. If you belong to a thyroid support group, there may be someone else in your area who would like to exercise with you.
- **Recognize your limits.** One of the main reasons people stop exercising is because they push themselves too hard to begin with and then become exhausted, injured or discouraged and give up altogether. It's essential to follow a proper, graded exercise programme. As you get stronger and feel more energetic, you will be able to take on more.

If you belong to a gym, the staff should be able to help devise a suitable regimen for you to follow and, in the UK, a GP can write an exercise prescription that allows you several free sessions at a local gym. If you can afford it, you might even want to consider a personal trainer to help you get started. Alternatively, there are plenty of good books and videos available as well as sites on the Internet.

• **Take advice.** If you are joining a gym or embarking on a formal exercise programme again after some time, always consult your doctor beforehand to check the suitability of your chosen activity. You should also explain your thyroid problems to your fitness trainer or instructor and describe how they affect you.

## Smoking

We're all aware that smoking is bad for the health, and it is becoming increasingly apparent that it is particularly harmful for individuals with thyroid disorders or any problems involving autoimmunity. Research shows that smoking is even more harmful to women than to men. Heart disease is a particular risk, given that thyroid sufferers are at an increased risk of cardiovascular problems, including raised cholesterol levels and high blood pressure – all good reasons to quit smoking.

Giving up smoking is often easier said than done. It is recognized that women in particular often smoke to relieve stress and avoid weight gain – this latter being more of a problem if you are hypothyroid. To stop smoking, you need to want to give up. The health benefits of stopping smoking are clear, especially if you have thyroid problems, and think of the money you would save. Setting aside the money you would have spent on cigarettes to treat yourself – to a holiday, a course of massage treatment or whatever – may be an added incentive to give up smoking. Giving up with your partner or a friend can also help maintain your motivation. You may need special help to identify why you carry on smoking and to find ways to stop. Your GP can tell you about smoking cessation clinics in your area, and may be willing to prescribe medications such as nicotine replacement therapy or Zyban to help you quit. Complementary therapies such as acupuncture and hypnotherapy may also help strengthen your resolve.

## Increasing Wellbeing

When you are feeling down, it's hard to feel that anything is going well. At such times, it may help to think about how you have dealt with adversity in the past and tell yourself, 'I got through that, I can get through this'.

Every day, make a list of things that have gone well or been rewarding. It doesn't have to anything major but, if you think about it, there's usually a little something – a hug from one of your children, a chat with a friend, someone who smiled at you, a magazine article or TV programme you enjoyed. Jan describes an encounter in the pharmacy that gave her renewed hope when she was feeling miserable:

> *I was standing waiting to be served and she came up to me and said, 'I know what's wrong with you. You've got Graves' disease'. She then said, 'I had eyes just like yours'. I looked at her and her eyes were perfect. It was just what I needed to enable me to believe that things would get better.*

Think of all the things that have upset you or made you angry, and allow yourself to let them go. Once you've got rid of any negativity, think of the next thing you are looking forward to – a cup of herbal tea, a workout or a walk, a book you want to read, a video or DVD you want to watch, something you plan to buy. Think about your environment – both outside and indoors – and how it affects your moods. Try to avoid leaving the curtains closed or the blinds pulled down during the day.

Put a bright cushion, picture or postcard of somewhere you've been or would like to go where you can see it. Be aware of sounds and the effect they have on you. Newspaper articles, TV programmes and song lyrics can all affect your sense of optimism and wellbeing. If you're feeling down, it's probably best to avoid reading gloomy news articles or watching programmes about death or disasters, or listening to gloomy songs. Instead, try listening to something relaxing or uplifting, such as dance music. Dance around your sitting room, or sing or hum along inside your head or even out loud.

## Your Relationships

Any change, especially illness, can affect close relationships, including that with your partner if you have one. We all tend to have a series of unspoken agreements with those around us. If something happens to disrupt these – for example, if you've always been active and independent or were the bright, cheerful, coping partner – the changes can put your relationships under severe strain.

Many relationships came under pressure as a result of the personality changes, physical changes in appearance and physical disability of thyroid problems. Changes in sex drive and activity as well as changes in the number of arguments you have with your partner (both common in thyroid disorders) are among the factors pinpointed as being stressful by psychologists in charts devised to quantify the impact of such life changes.

The best way to avoid problems is to keep the lines of communication open and let others know how you feel. Try not to talk only about the times you are feeling tired and lacklustre, but also about the days when you are feeling better.

Men in general can have more problems dealing with illness than women. Your partner may be experiencing feelings of weakness or fear which may, in turn, threaten his sense of competence. He may feel protective towards you and experience a sense of failure when your thyroid problem reveals to him his powerlessness.

- Give yourself time. Recognize that it takes time to adjust to any new situation.
- Talk to others about your condition and the many ways it affects you. However, try not to let your thyroid dominate your life, especially once problems have been sorted out. Do things you enjoy with your friends and family, such as going for a meal, a walk, to the cinema or on holiday, and forget about your illness for a while.
- Be sensitive to the personality style of those around you. Recognize that their way of dealing with issues will not always be the same as yours, so be tolerant. Play to your loved ones' strengths. One friend may be happy to spend hours on the phone discussing the medical nitty-gritty while another may be happier to go out with you or pay you a visit. Don't expect everyone to respond in the same way.
- If your partner is unsympathetic, find a sympathetic ear among your friends or seek out a counsellor or therapist to help you deal with your feelings.
- Recognize that every relationship has its ups and downs. Don't blow these out of proportion, and try to believe in your ability to pull through. Try to avoid minor irritants becoming major issues. Talk about them and try to be patient.

## Looking After Yourself

Developing a chronic illness can offer an opportunity to look at your life and the way you live it. Many women rush around caring for others and putting their own needs last. They skip meals, miss out on vital sleep, and neglect their own physical and emotional needs – all of which results in stress. We've seen how the nervous and endocrine systems release hormones and other biochemicals that disrupt the function of the immune system, and how this may be implicated in thyroid autoimmunity and other illnesses. Chronic illnesses such as thyroid disease can make it more difficult to deal with stress, and create a vicious circle in which stress increases the symptoms that make you feel more stressed.

It's impossible to eliminate stress altogether, but you can learn to control it by being alert to warning signs such as anxiety, irritability, mood swings and fatigue at an early stage. You may find it helpful to keep a stress diary to identify what it is that you find stressful. This will enable you to avoid people or events that stress you and allow you to plan how to deal with them. It's important to look after yourself physically – to eat a healthy diet and exercise, and do things for yourself that you enjoy, such as making time to take a walk, have a massage, read a book or phone a friend.

Sabeha, who developed hypothyroidism after the birth of her son Louis, describes how looking after herself made her feel better:

> *I was feeling pretty awful despite the change in*
> *my medication, so I decided to take matters into*
> *my own hands. I've cut right down on alcohol.*
> *I'm just having two or three units a week now.*
> *I'm trying to follow a good diet and eating*
> *regularly, and I'm not feeling so tired. I'm also*
> *exercising. I got a personal trainer and I'm doing*
> *a mix of weights and the treadmill. Although*
> *I'm still about a stone overweight, I feel like I'm*
> *toning up, my clothes fit better and I feel generally*
> *healthier.*

## Treat Yourself Better

The lack of self-esteem that accompanies illness can make it hard to give to yourself. At the same time, if you feel physically unattractive, you may feel you don't want to bother with clothes or make-up. Nevertheless, looking after yourself and pampering yourself in small ways can make a real difference to how you feel as well as help you improve your appearance which, in turn, can make you feel better. If you catch yourself feeling guilty about spending money or time on yourself, tell yourself you're worth it. Even something as small as buying a beautiful scarf to hide a goitre or the scar on your neck can do wonders for your self-esteem.

- **Look after your hair.** If your hair is thinning, the number of 'bad hair' days may exceed the good days. So be sure to go for regular haircuts, and wash and condition your hair regularly. Improving your diet and taking exercise can help strengthen hair and reduce any hair loss. There is also a vast number of products on the market that can thicken your hair and improve its condition. Colouring your hair can add volume, too, as can a good cut or a root perm. Incidentally, thinning hair tends to look fuller if you keep it short.
- **Get ahead, buy a hat.** Make wearing a hat or scarf part of your personal style. Alternatively, think about investing in a wig. There are some very convincing ones available so no one ever needs to know you're wearing one.
- **Master the art of disguise.** There's a fantastic range of cosmetics available today, from natural-looking fake tans to light-reflecting foundations that can help your skin look healthier and disguise imperfections. If your skin is dry, invest in a good moisturizer. A visit to a beauty salon or the cosmetics counter of your local department store can give you new ideas about make-up and how you can look better. Attractive clothes can help you make the best of your figure even if you are over- or underweight.
- **Watch your eyes.** Even if you don't have thyroid eye disease, you may want to be careful about how you treat your eyes. Use a hypoallergenic product that won't irritate your lids. If you have sparse eyebrows, use an eyebrow pencil to make them look thicker, or consider a eyebrow tint.
- **Treat yourself to a massage.** As well as soothing away stress and tension, allowing someone to touch you can

help you accept yourself, and make you feel better about your body and the way you look. A facial massage can help disperse excess fluid around the eyes and rid the tissues of toxins.

- **Go easy on yourself emotionally.** Avoid phrases like 'I should', 'I ought' or 'I must'. Instead, allow yourself to make choices with words like 'I could'. You may not be exactly the same as you were before you developed thyroid problems, but life can still be good.
- **Treat yourself.** Do something new for yourself each week. Buy yourself a bunch of flowers, wear a new perfume, take a walk in the park or go to the movies. Take time out to exercise or relax, or try some meditation.

If you've just been diagnosed with a thyroid disorder, it may feel as though you'll never get better. However, all of the people quoted in this book have come to terms with their thyroid problems and now enjoy life – and so can you.

# CHAPTER SEVEN

# Integrated Treatment: Complementary Therapies

Conventional medicine has a great deal to offer to help you manage your thyroid. But, as we've seen, it doesn't solve every problem, and many of those with thyroid problems claim that complementary therapies have helped them deal with consequences of thyroid disease that weren't addressed by conventional medicine. Therapies like acupuncture, aromatherapy, homoeopathy, massage, nutritional therapies, reflexology and yoga can help you relax and often alleviate the recalcitrant problems – such as depression and anxiety, digestive disorders, fatigue, insomnia and weight gain – that so often go hand-in-hand with thyroid disease.

In the past, complementary therapies were often regarded with suspicion by the orthodox medical profession. However, this is changing as more and more doctors come to recognize that such therapies can be a useful adjunct to conventional treatments and help symptoms that are not easily treated by conventional medication or therapy. As we move more and more towards a pick-and-mix style of medicine, complementary therapies do have a place in what a number of doctors now call 'integrated medicine', which recognizes that the environment in which you live, the food you eat, the relationships you have with those around you, and your own mind, body and – yes – spirit all play a part in healing and how well you feel.

This is not to dismiss the effective, orthodox medical treatment available for thyroid problems. You may well feel a new woman once the appropriate treatment has been found for your thyroid. However, as with many chronic, long-standing illnesses, thyroid problems can take months or even years to sort out. In the meantime, your symptoms may, at best, only be alleviated or prevented from progressing. At worst, there may be relapses, side-effects and little or no improvement. Your doctor may become tired of hearing your tale of unremitting woe, and you may become frustrated that he doesn't appear to listen or take your problems seriously.

Donna Beckwith, writing in *BTF News* some years ago, echoes the feelings of many women with thyroid disorders when she says:

> *Although looked at in clinical medical terms*
> *thyroid disease is basically straightforward and*
> *fairly easy to treat, often the reality of the*
> *symptoms during times of imbalance are much*
> *more difficult to live with and can be very*
> *distressing ... Knowing your blood levels are OK*
> *doesn't help much when you're feeling dreadful.*

This is where complementary therapies can be a real help. Good healers of every persuasion have always paid attention to more than simply the relief of symptoms. The existence of the 'placebo effect' – the phenomenon in which an illness improves even with a non-active placebo treatment – is proof of the power of mind over body. In complementary medicine, the placebo effect is a recognized and important part of the healing rather than dismissed as an irrelevancy, as it sometimes is by conventionally trained doctors. What complementary therapies do is acknowledge that the influences on wellbeing don't stop at the door of the doctor's surgery, and that there are many ways to encourage health and healing.

## A Whole Person

What complementary therapies have as an underlying theme is that they treat the patient as a whole person. Such an approach is particularly relevant for those with thyroid problems, where the effects of the condition are very wide-ranging and have knock-on effects on so many other aspects of life. Furthermore, complementary therapies may be particularly helpful for thyroid problems as they are often extremely relaxing and, as we have seen, symptoms are frequently brought on or exacerbated by stress.

## Time to Talk

While the average visit to a conventional doctor is usually a speedy affair with much of it spent scribbling on your notes or writing a prescription, a visit to a complementary practitioner is likely to include more time for talking and discussing things that may seem unrelated to your thyroid. This generally includes details of your work, home life, diet, state of mind, relationships, free-time pursuits, your health in the past and the health of the rest of your family. The aim of this information collecting is to give the practitioner a more complete picture of you as a person.

## Partners in Healing

Although it is always preferable for you to become an active participant in your own treatment with your doctor, because conventional medicine is so dependent on complex drugs and technology, it isn't always easy to feel you are an equal partner. In contrast, complementary therapies tend to offer simple ways of encouraging healing. Just visiting a complementary practitioner can make you feel as if you're taking control of your condition. What's more, many complementary therapies demand some involvement from you – it's not just a question of taking the tablets and going for blood tests. You may be asked to actively change your diet, exercise habits and other aspects of your lifestyle. This can be tremendously empowering, especially when you have a condition in which your body sometimes seems to be running you rather than the other way around.

## A Matter of Balance

Another feature that many complementary therapies have in common is that they are concerned with balance, and it should be clear from this book that thyroid problems are all to do with balance. According to any number of non-conventional therapies, illness is the result of an imbalance and the body's attempts to right itself. It's not difficult to see how this concept can be easily applied to thyroid disease.

In herbal medicine, for example, the ancient concepts of 'hot' and 'cold' disease characteristics fit in perfectly with the insights of 20th-century endocrinologists and thyroid disease. Individuals with a 'hot' condition are likely to have a high metabolic rate and feel hot, stressed and nervous. People with a 'cold' condition, on the other hand, will feel cold, have a sluggish metabolism and tend to be overweight.

Acupuncture, too, views all illness as a disturbance in the balance of two opposing qualities – in this case, *yin* and *yang* (*see page 151*). Although the endocrine system is not recognized as such in Chinese medicine, there are a few studies showing that acupuncture may be successful in treating disorders in which there is a hormonal imbalance, such as subfertility.

## Which Therapy?

So, which complementary therapies are likely to be the most beneficial? As thyroid problems have such an impact on the mind as well as the body, therapies that act on the body through the mind, and vice versa, should be particularly helpful. The therapies listed below are only a few suggestions to start you thinking. When it comes to choosing a therapy, it's very much a matter of experimenting and finding the system, or even the practitioner, that suits you personally, as complementary therapies, much more than conventional ones, are based on treating you as an individual. And, as many of these therapies work better if you're in a positive frame of mind, using a therapy that appeals to you is a powerful part of the treatment. So, if you are drawn to a particular complementary therapy, apply a little trial and error until you find the one that works.

## Acupuncture

Acupuncture is very much about rebalancing energy levels. According to Chinese medicine, disease occurs when the body's internal balance is disrupted in some way. Harmony and balance depend on the smooth flow of the body's life force, or energy, known in Chinese medicine as *ch'i*. This circulates along invisible pathways or channels. The smooth flow of energy also depends on the correct balance of two qualities known as *yin* and *yang* in the body.

*Yin*, the feminine principle, is associated with coldness, darkness, wetness and softness, while *yang* is linked with the more masculine qualities of hardness, brightness, heat and dryness. Illness can be due to excessive *yin*, bringing about pallor, cold hands and feet, and depression or, alternatively, a shortage of *yang*, with tiredness and poor circulation – in fact, all the symptoms of an underactive thyroid. Illness may also be due to a shortage of *yin*, leading to insomnia, dry mouth and nervous excitement, or an excess of *yang*, characterized by heat and overactivity, a flushed face, anger, anxiety and stress – symptoms that correspond remarkably to an overactive thyroid.

DIAGNOSIS AND TREATMENT

The diagnosis is made after the practitioner has examined you for signs of the various disturbances and taken several pulses. According to Chinese medicine, there are six pulses – three on each wrist. Each pulse has three depths and as many as 28 different pulse qualities. Checking these pulses enables the acupuncturist to come to a very detailed diagnosis of your condition.

The acupuncturist also examines your tongue, making a note of its colour, shape, thickness, moistness, and the presence and location of any coating. The practitioner may also perform a physical examination.

The acupuncturist then inserts sterilised needles at specific points along the body's meridians to restore the body's normal balance of *yin* and *yang* and strengthen *ch'i*. The meridians associated with the large intestine (in the neck area) and the gallbladder are considered especially important for thyroid problems. You may feel a slight pinprick as the needles are inserted or, occasionally, you may feel a tingling sensation related to the position of the needle.

Practitioners also use the dried leaves of the herb mugwort, which are burned and placed around the needles to warm and tone the *ch'i* when the condition is characterized by cold and damp. This technique is known as moxibustion. Alternatively, a moxa stick (of burning herb) may be held close to the acupuncture point.

## Aromatherapy

Aromatherapy, the use of essential oils from plants and herbs, is becoming increasingly popular as a safe, gentle and relaxing therapy. It is often particularly helpful for alleviating stress and easing symptoms such as depression, anxiety and insomnia. The oils can be added to your bath, inhaled or used with a carrier oil in massage. Different oils are reputed to have different properties. Lavender and clary sage, bergamot and jasmine are said to be calming whereas cinnamon, ginger, peppermint, rosemary and pine are thought to be stimulating, and jasmine, lavender, neroli, rose and ylang-ylang considered uplifting. You can use these oils for self-help to help increase your energy or calm you down when tiredness or anxiety is the result of your thyroid problems.

Aromatherapists believe that the oils are absorbed into the bloodstream and have biochemical effects on the body. Orthodox doctors tend to be rather sceptical about this, but acknowledge that a therapeutic massage can help ease pain and promote relaxation.

## Autogenic Therapy

This is one of the few complementary therapies studied specifically in relation to thyroid problems. Developed in the 1920s by German doctor Wolfgang Schultz, autogenic therapy involves a series of exercises designed to focus the mind by asking you to concentrate on feelings of heaviness, warmth in the limbs, a calm heartbeat, easy natural breathing, abdominal warmth and cooling of the forehead. You are to practise the exercises while sitting comfortably or lying down, three times a day after meals, for about 10 minutes each time.

Mild anxiety states respond particularly well to this therapy. After four or five weeks, the patient can be gradually weaned off whatever tranquillizers, beta-blockers, sleeping pills and other medications are being taken to treat anxiety.

Autogenic therapy draws on some of the insights of meditation. As you focus inwardly, the stresses and strains of daily life are left to one side for a time so that the body's own healing and relaxation abilities can be called upon to restore you. The various techniques, learned initially by attending a course, have been found to lower heart rate, blood pressure and improve emotional balance.

Often, buried feelings of anger, grief or anxiety surface during the training. Such reactions are normal and proof that the therapy is working. The therapist can help you deal with these strong emotions, which can be enormously healing.

Research has shown that a number of typical symptoms of an overactive thyroid, such as sweating, tremor, nausea, vomiting, diarrhoea and irritability, diminish gradually over the course of autogenic therapy.

## Herbal Medicine

Herbal medicine has been used successfully for centuries to treat illness and cure disease. Today, some eight out of 10 of the world's population still rely on herbs as medicine. Herbs and plants contain many active, medicinal ingredients in their bark, seeds, leaves and flowers. Indeed, many medical drugs were synthesized originally from herbs. Aspirin derives from white willow, the heart drug digitalis from foxglove, and taxol, used to treat advanced breast cancer, is based on the leaves of the yew tree.

Herbs can be powerful inflammation fighters, which makes them useful for treating autoimmune thyroid conditions that cause inflammation. However, their very effectiveness makes it important that they be used with caution. Because the chemicals in herbs are pharmacologically active – they alter the body's chemistry – you should always consult a trained, qualified and experienced herbal practitioner. You should also inform the doctor who is treating your thyroid problem that you are planning to use herbal treatment as there have been reports of herbal remedies interacting with conventional thyroid drugs. For example, celery seed extract (*Apium graveolens*), used to treat arthritis, fluid retention and cystitis, may lower levels of thyroxine ($T_4$). Other reports suggest that $T_4$ may interact with St John's wort, a popular treatment for mild-to-moderate depression. This is not to say you shouldn't use herbal remedies, but it does underline the importance of making sure you inform your doctor and the pharmacist when you are picking up your prescription that you are using other medicines.

## Homoeopathy

Like many other complementary therapies, good health to a homoeopath is a state of balance and disease is a result of a weakening of the body's vital force or energy. Homoeopathic treatment is aimed at strengthening and nourishing this vital force to help the body heal itself.

Remedies are based on the idea that a substance that produces certain symptoms when given to a healthy person can be given to a sick person with the same symptoms and restore health. Homoeopathic remedies are made from plants, herbs, minerals and other substances that have been repeatedly diluted and shaken or succussed – a process said to increase the power of the substance, or potentize it. As homoeopathy is aimed at a person's vital force, so the pure energy of the remedy is said to stimulate the weakened vital force. And because it is of the same nature, it is able to stimulate and nourish the vital force, which is then able to perform its job properly and restore the body to harmony and health.

Practitioners use what is called the law of similars to devise various symptomatic pictures, which enable them to choose the right remedy. For example, *Belladonna* patients have wild, staring eyes, talk too fast, are impatient, bad-tempered and irritable, and can't stand direct sunshine – characteristic symptoms of an overactive thyroid.

To a homoeopath, every person is different, and the same symptoms in two different people could have different diagnoses and treatments. The symptoms of scarlet fever in one patient may point to a prescription of *Belladonna*, yet another patient with scarlet fever may present with a different set of symptoms and, hence, require a different remedy. Because of this, diagnosis is an extremely skilled business for all but the simplest conditions. If the idea of homoeopathy appeals to you, you should visit a qualified practitioner. Some homoeopaths are also trained in orthodox medicine and it may be beneficial to visit such a doctor, who can then monitor your progress with conventional drugs while prescribing homoeopathic remedies.

## Nutritional Therapies

Nutritional therapies encompass a number of different treatments in which what you eat is considered the foundation of good health. There's little doubt that eating a healthy diet containing plenty of fresh fruit and vegetables can help increase energy, and improve physical and mental health. Many nutritional practitioners, however, go further by arguing that, even with a healthy diet, it is possible to be deficient in vital nutrients because of the poor quality of soil, use of pesticides and long storage time for foods, and because infection, ageing and certain drugs can affect how well the body absorbs the nutrients in food.

For these reasons, many nutritional therapists recommend taking nutritional supplements – vitamins, minerals and other nutrients – to help bolster health. But nutritional supplements, like herbs, can affect body chemistry and, although a good multivitamin/mineral supplement shouldn't hurt you, it may be wiser to take any other supplements only under the supervision of a practitioner who is experienced in nutritional therapy as supplements can affect your body in the same way that drugs can, and may interact with other treatments for your thyroid problems.

### A WORD ABOUT KELP

Kelp tablets are made from seaweed and contain iodine. In persons without thyroid problems, they are said to balance thyroid activity, but the story is different for those with a thyroid disorder as the amount of iodine in kelp preparations cannot be as easily measured as it can from other sources.

In general, iodine deficiency is no longer a problem for most of us in the West as many foods are fortified with iodine. Sea fish is also a good source. Kelp, however, should be used cautiously by anyone with a thyroid problem as it can provoke hyperthyroidism. It is best, therefore, not to dose yourself with kelp, and to only take it on the advice of a qualified herbal or nutritional practitioner.

Other foods, known as goitrogens because they interfere with iodine absorption in the thyroid, can worsen symptoms if you have an underactive thyroid. Goitrogens are most often found in plants of the brassica family, including cabbages and turnips, as well as peanuts, soya, pinenuts and millet. In *The Natural Health Handbook*, nutritionist Dr Marilyn Granville observes that these foods seem to be a problem mainly when eaten raw and in excess. Goitrogens may help lower excessive thyroid production in women with hyperthyroidism. Dr Granville's book includes a hormone-balancing diet for people with minor hormone imbalances. However, it's important to have a proper diagnosis and work with a conventional doctor and nutritional therapist to find the treatment that is best for you.

## Reflexology

Reflexology, which involves foot and sometimes hand massage, is another therapy that uses the concept of *ch'i* or energy flow. According to reflexology teaching, the body has 10 channels, beginning or ending in the toes and extending to the fingers and the top of the head. Each of these channels is associated with a particular organ of the body. The area concerned with the thyroid is at the base of the big toe, with two more points on the pads of the feet beneath and between the big toe and the toe next to it.

Minute crystalline deposits are said to form in areas where energy is blocked. By massaging these points, the therapist aims to correct the energy flow and restore balance. Thyroid problems are said to be helped by massaging the reflexology areas relating to the thyroid as well as those linked to the adrenals, pituitary and ovaries. The massage is usually gentle, but it may be momentarily uncomfortable if the practitioner presses deep into the foot.

'Glandular disease' is among the illnesses listed by Eunice D. Ingham, one of the original proponents of reflexology, in the 1930s. However, even without the potential therapeutic benefits, you may well find this therapy tremendously relaxing and calming.

## Yoga

Yoga is a complete system of philosophy that is part of the ancient Indian medical system of Ayurveda. The word, which is Sanskrit for 'union', aims to unite mental and physical health. In more modern times, yoga has been used successfully by orthodox doctors to help treat a number of disorders. A US heart doctor, Dean Ornish, uses it to reduce the symptoms of heart disease.

There are many different types of yoga, the most common in the West being hatha yoga, which concentrates on calming the mind through the use of physical postures – *asanas* – and breathing techniques – *pranayama*. The currently popular forms of yoga, such as astanga 'power' yoga or Iyengar yoga are all types of hatha yoga. Interest is increasing in the use of specific *asanas* to treat certain conditions.

Yoga can help ease stress, anxiety, moderate depression and menstrual problems. You should always tell the yoga teacher if you have any medical problems as some *asanas* may then not be advisable. Yoga is a very gentle form of activity, and you will always be encouraged to work at your own level and increase gradually as you become more adept. You're unlikely to be expected to do difficult postures like headstands at first – or even ever!

## Finding a Therapist

As thyroid problems are so complicated, self-treatment is not recommended. However, choosing a complementary therapist can be tricky as, in some cases, there are no standard qualifications as there are for orthodox practitioners. By personal recommendation is one way to find a good therapist. Alternatively, you may want to approach one of the umbrella organizations that maintains the standards for various alternative and complementary therapies, or you could ask your doctor. Increasing numbers of orthodox physicians are now taking such alternative therapies seriously and some even allow complementary practitioners to practice on their surgery premises.

## Watchpoints

As complementary treatments are sometimes perceived as being more natural than conventional treatments, it is easy to imagine they are always safe. The truth of the matter is that any effective treatment – complementary or conventional – involves both benefits and risks. One of the hazards in complementary medicine is that a serious illness may go undiagnosed and untreated. Also, as we've seen, there is a risk that some treatments or remedies may interact adversely with conventional drugs or treatments, or be dangerous in their own right. To avoid these possible dangers, it may be wise to take a few precautionary steps:

- **Don't rely exclusively on complementary treatment.** Make sure you get a proper diagnosis of your thyroid problem before consulting a complementary practitioner. It would be unwise to place your trust exclusively in the hands of a complementary practitioner as proper medical management is necessary to restore thyroid function.
- **Don't stop taking your medication.** If you've been prescribed medication for your thyroid problem, don't abandon it. Certain drugs, especially when you stop taking them suddenly, can have serious side-effects. Steer clear of any practitioner who encourages you to give up a proven effective medication for one that is unproven.
- **Consult a competent practitioner.** Make sure anyone you consult belongs to a professional therapy organization (*see page 233*).

- **Let your doctor know.** If you decide to visit a complementary therapist, you should tell your doctor that you are doing so. A few doctors may be resistant to the idea of complementary medicine but, as mentioned, many others are beginning to recognize that complementary therapies can reach the parts that conventional medicine doesn't.
- **Check it out beforehand.** Before going ahead with a complementary treatment, check the consulting rooms to make sure they are clean and hygienic. If you are consulting an acupuncturist, check that he uses disposable, sterile needles. If you are treating yourself with complementary remedies, check that they are properly sealed and packaged before you buy them, and follow any instructions given with them.

There are some situations in which special caution is needed:

- **Before an operation.** Some herbal medicines can affect blood-clotting or alter the way anaesthetics work. If you are taking herbal remedies, inform the surgeon and stop taking them a few weeks before the operation.
- **If you are pregnant.** The fetus is particularly at risk from harmful substances that may cross the placenta, so it's advisable to take as few medications as possible during pregnancy, including herbal remedies and food supplements. Check with your doctor or midwife and only take such remedies on the advice of a trained complementary practitioner.
- **If you are over 60.** The body often becomes less efficient in dealing with medications, including herbal remedies and supplements, as we get older. Check with your doctor and complementary practitioner, and be sure to note any side-effects.

Just as with conventional therapy, it is wise to keep an open mind and not accept everything you are told, lock, stock and barrel. If a therapy isn't helping after a reasonable period of time, you don't have to carry on using it, whatever anyone says. Likewise, if you develop new symptoms, it's important to seek a proper medical diagnosis. Don't simply accept it as a sign that the treatment is working, as some may suggest, as there is the possibility that an important or significant factor has been missed.

Keeping these important provisos in mind, complementary therapies nevertheless can be a wonderful way to help ease yourself back to wellness and can be useful additions to conventional medical treatment.

# CHAPTER EIGHT

# The Eyes Have It: Thyroid Eye Disease

One of the most cruel and puzzling conditions associated with thyroid disorders is thyroid eye disease, which is autoimmune in origin and estimated to affect some 250,000 people in the UK. Often, but not always, it is linked to Graves' disease and is sometimes called Graves' ophthalmopathy. This term is, however, misleading because thyroid eye disease can also affect persons with hypothyroidism and even – in 10–15 per cent of cases – those with normal thyroid function.

Dr Nancy Patterson, founder and President of the US National Graves' Disease Foundation, gives a poignant, personal account of how the eyes can be affected on the Thyroid Federation International's website:

> *My eyes were an immediate target of the misguided antibodies that ravaged every cell they could find. My expressive, dancing dark-brown eyes and my naturally curly hair were my two favourite features. They were among the first casualties of the chaos of Graves' disease.*

## Who Is At Risk?

Up to half of all patients with Graves' develop thyroid eye disease although, curiously, eye symptoms may become evident well before or after the onset of the thyroid disorder. The advent of more sophisticated scanning techniques has revealed that, in fact, Graves' disease has some effect on the eyes of virtually all of those affected, even though it may not be immediately – or ever – apparent.

The good news is that, over the past few years, a great deal more is now known of the potential causes of thyroid eye disease, and with this new understanding have come improved medical and surgical ways of treating its symptoms. The bad news is that, sadly, no way has yet been found to prevent it altogether.

## Symptoms and Signs

Although many people associate thyroid problems with bulging eyeballs, known as proptosis or exophthalmos in the medical jargon, in fact, these disorders can affect individuals in many different ways – and not just the eyes. Writing in the *TED Newsletter*, British endocrinologist Dr Malcolm Prentice describes three groups of symptoms. First, there are those arising from an overactive or underactive thyroid and, second, there are those affecting the eyes. Symptoms may include:

- prominent bulging eyes (proptosis or exophthalmos)
- puffiness around the eyes and marked bags beneath the eyes
- dry eyes caused by an inability to close the eye properly
- watery eyes
- soreness and redness
- a sense of pressure within the eye socket
- burning and 'grittiness' in the eyes
- pain or discomfort when looking up, down or sideways
- light intolerance (photophobia)
- double vision (diplopia), caused by weakness of the eye muscles (ophthalmoplegia), which may come and go at different times of the day

- blurred vision and/or loss of colour perception, caused by pressure on the optic nerve
- lid lag, when the upper lids are slow to follow the downward movement of the eyes when looking down
- lid retraction, when the upper eyelids are pulled upwards to expose more of the whites of the eyes than normal, producing the effect of a constant stare that may be confused with proptosis.

Symptoms can affect one or both eyes and may be particularly troublesome at night, in air-conditioned buildings, in places with hot-air heating and on windy days.

The third group of symptoms identified by Dr Prentice comprises those experienced by persons who have normal thyroid function and thyroid eye disease. These include non-specific symptoms such as aches and pains, fatigue, depression and a general feeling of unwellness.

Symptoms of thyroid eye disease may emerge at different times. In women with Graves' disease, the eye symptoms appear before the symptoms of hyperthyroidism in about 20 per cent of cases, at the same time in 40 per cent of cases and after the onset of symptoms in 40 per cent of cases. Lucy's story is fairly typical:

> I have Graves' disease, which presented as
> diarrhoea. At first, the doctors thought there was
> something wrong with my gut, and I was referred
> to the hospital for tests. Meantime, my eyes felt
> really sore and gritty. I was given steroids for the
> gut problems, so the eyes didn't bother me for a
> while, but eventually, when I did get diagnosed
> with Graves', I stopped taking them, and then my
> eyes did start to really play up. At first, they were
> very sore and red, and other people noticed they
> were changing in appearance, though I was
> unaware of it to begin with. They became very
> sore and swollen with an incredible feeling of

*pressure. Then I started to find it hard to move the muscles, and now I have to turn my whole head to look left and right. My upper lids are very swollen and puffy. I had some lid lag and straining and had horrible bags under my eyes, though that is better now.*

---

### Table 8.1
### Signs and symptoms of thyroid eye disease

| What you may experience | What others may notice | What your doctor may detect |
|---|---|---|
| Painful red, watery, itchy eyes | You look different and may even be unrecognizable | Oedema (swelling due to fluid accumulation) of the eyelid |
| Soreness and aching behind the eyes | Your eyes appear to bulge and stare | Red eyelids |
| Blurred or double vision, especially when looking from side to side at certain times of day | You look as if you've been crying | Swelling (chemosis) of the conjunctiva (membranes covering the whites of the eye and lining the inside of the lids) |
| Can't go out without sunglasses (photophobia) | You're always wiping your eyes | Red eyes or bright red blood vessels that are normally not visible |
| Sensitivity to dust and wind | You look as though you've had too many late nights | |
| Can't wear contact lenses (as a result of dry eyes) | | |

---

## How Thyroid Eye Disease May Affect Your Life

Some of the worst consequences of thyroid eye disease are the effects it can have on your confidence and self-esteem. As one member of the UK's Thyroid Eye Disease Association told me:

*You feel awful, people stare at you, children make comments. I wouldn't go out for 12 months. It*

*isn't life-threatening, but it affects your whole life.*
*Some people become suicidal.*

The Thyroid Federation International website describes how, for some people, relationships may break down, jobs are lost and some become prisoners in their own homes unable to face the world outside. As Jan, who developed thyroid eye disease with Graves' disease, recalls:

> *I didn't really notice it myself because I don't look in the mirror much, but other people kept commenting on it. I looked like a gargoyle with horrible bulgy eyes and virtually no hair. People thought I was on drugs. Eventually, I became so self-conscious that I avoided going out.*

Lucy, another woman with thyroid eye disease, comments:

> *It is a horrible thing to happen. One of the worst things was that people who hadn't seen me for some time couldn't recognize me.*

It's not just a changed appearance either. Thyroid eye disease can make it difficult or impossible to read and may affect depth perception, which can have an effect on walking up or down the stairs, while overlapping numbers due to double vision can lead to errors in adding up figures or doing mathemetical tasks. According to Thyroid Federation International, many 'literally run into the wrong wall'.

## What Causes Thyroid Eye Disease?

Thyroid eye disease is a result of the body's immune system attacking itself. The main targets are the orbital fibroblasts, the cells lining the pear-shaped bony sockets, known as the orbits, that lie behind the eye. The orbits are lined with protective pads of fat, connective tissue, blood vessels, muscles, nerves and the lacrimal glands, where tears are formed. Inflammation causes the tissues to thicken and triggers the release of cell proteins called cytokines, which cause further inflammation. This, in turn, leads to the release of other chemicals that attract fluid into the orbit and cause further tissue swelling.

As with other types of autoimmune disease, it's thought that a gene may be involved in determining both the susceptibility to thyroid eye disease and its severity. Scientists all over the world have been searching for likely candidates, and a possible culprit has been identified by a team from the UK's Newcastle-upon-Tyne University. Writing in the *BTF News* (Summer, 2001), Dr Simon Pearce, of the Endocrine Genetics Research Group, reported the discovery of a variation within a gene known as CTLA4, which lies on the long arm of chromosome 2. Intriguingly, the gene is in the same position as a previously located gene for Graves' disease.

The researchers were astonished to find the mutated gene in six out of 10 people with the most severe form of thyroid eye disease. As Pearce observed, 'This was a particularly exciting finding, which has since been confirmed by a study in Italian patients, because previously the most important influence on the development of thyroid eye disease was thought to be environmental.'

As with other autoimmune thyroid diseases, researchers suspect that a host of mainly as yet unidentified agents may be involved in triggering the autoimmune activity, but one already known trigger is smoking. As long ago as 1993, an article in the *Journal of the American Medical Association* showed that not only were smokers twice as likely as non-smokers to develop thyroid eye disease, but they were also more likely to be severely affected. More recent research has found that smoking can also diminish the effectiveness of treatments for the condition. A study published in the *Annals of Internal Medicine* in 1998 showed that treatment was four times more likely to be effective in combating eye problems in non-smokers than smokers. The same study showed that mild eye problems were also more likely to progress and become severe in those who smoke.

## Course of the Disease

Thyroid eye disease goes through an active phase followed by an inactive (burnt out) phase. For most patients, the symptoms are mild and last only a relatively short time – up to a couple of years – before they fade or disappear. But, in about one in 10 cases, the inflammation persists and symptoms get worse. If this is the case, it will usually happen within a few months of developing the symptoms. If your eyes have remained the same for more than six months, it would be unusual for them to get worse. While the disease is active, as with other autoimmune diseases, there may be periods of increased disease activity (flare-ups) interspersed with periods of relative calm.

If your eyes are only mildly affected, there's a fair chance they will return to normal over the course of a year or so. If they are more severely affected, there may be permanent changes in appearance caused by the inflammatory process which can, over time, lead to the development of fibrous scar tissue.

## Getting a Diagnosis

As with other thyroid problems, it can be difficult to get a diagnosis. Because thyroid eye disease is relatively rare, symptoms can be mistaken for other problems, such as hayfever or conjunctivitis. As Janis Hickey, chair of the British Thyroid Foundation, writing some time ago in the *BTF News* recalled:

> *I remember waking up one morning ... and*
> *noticed that my eyes looked and felt peculiar. One*
> *GP thought it was conjunctivitis and prescribed*
> *eye drops. When I returned a few weeks later, as*
> *there had been no improvement, the GP's response*

> *was that conjunctivitis took some time to clear up.*
> *I had very bad double vision by the time advice*
> *was sought of another GP.*

After the birth of her first baby, Janis had another bout of eye problems:

> *My vision was bad again and when I quizzed a GP*
> *about this, he said, wrongly, that it had nothing to*
> *do with my thyroid disorder. Consequently, I went*
> *for an eye test and, despite alerting the optometrist*
> *to the thyroid disorder, was prescribed a pair of*
> *glasses that were of little use; my eyes were so*
> *prominent by now that my eyelashes touched the*
> *lenses. Over a period of some months, I suffered*
> *from raging headaches and pain behind the eyes.*

Since the doctor will base diagnosis on the appearance of your eyes (*see page 163*) and your description of your symptoms, it's important to be as specific as you can when describing symptoms. Dr Malcolm Prentice advises, 'Rather than saying "my eyes feel terrible", try to analyze symptoms, such as grittiness in the eyes, pain or discomfort of the lids, or double vision in a certain direction which comes and goes at certain times of day.'

The doctor will supplement what you say with an eye examination, which may involve measuring the degree of protrusion, swelling and any deficiency of eye movements. You may be referred to an ophthalmologist for more detailed tests and examination. Blood tests may be performed to look for thyroid-stimulating antibodies – present in nine out of 10 persons with thyroid eye disease – and to check for lowered levels of thyroid-stimulating hormone (TSH).

In some instances, more detailed investigations such as specialized magnetic resonance imaging (MRI) or ultrasound scans may be needed to assess eye muscle size and swelling to evaluate the extent of inflammation.

If thyroid eye disease is suspected, ideally you should be referred to a unit with access to both an endocrinologist and ophthalmologist with experience of the diagnosis and treatment of thyroid eye disease, as well as other specialists such as an orthoptist (someone trained in how the eyes move and work together as a pair). As there are only a handful of such units in the UK, you may have to do some investigating to find the one nearest you.

## Clinical Activity Score

Like many autoimmune diseases, thyroid eye disease can have periods of increased disease activity alternating with periods of relative calm, when inflammation is dampened down and symptoms such as redness, pain and swelling are less troublesome. The disease may be severe – with pronounced symptoms such as very bulging eyes – yet, at the same time, be inactive – with no signs of active inflammation.

An important part of deciding upon the most appropriate treatment is to assess the severity of the disease and how active it is. The clinical activity score is a way to quantify disease activity that uses a special scoring chart to check off signs such as pain, redness of the eyelids, eyelid swelling and other classic features of inflammation. A score of three or more suggests that the inflammatory process is active.

## Treatment Options

The treatment of thyroid eye disease is complex and dependent on how active and severe the disease is. In mild cases, no medical treatment may be needed, although practical measures such as eyedrops to soothe dry eyes and plastic prism lenses to help improve double vision (*see Table 8.2*) may be advised.

# Table 8.2
# Treatment options* for thyroid eye disease

| Symptom | Possible treatments |
|---|---|
| Irritation and redness of the eyes | • Artificial tears/'comfort drops' from an optometrist or pharmacy to help keep the eyes moist<br>• Soothing ointments such as Lacritube |
| Puffiness around the eyes | • Extra pillows/a bolster to raise the head while sleeping or raise the head of the bed to encourage excess fluid to drain away<br>• Diuretics (water tablets) to increase the amount of fluid collected in the urine<br>• Surgery if severe puffiness persists, once the disease is inactive |
| Staring eyes | • Tinted spectacles to disguise the eyes<br>• Surgery, if severe, once the disease is inactive |
| Double vision (diplopia)† | • Plastic prism lenses, or thin pieces of plastic stuck onto glasses with water or semiopaque surgical tape<br>• Watching TV wearing a patch over one eye<br>• Immunosuppressive drugs if it gets worse (see below<br>• Surgery to realign eyes |
| Deteriorating vision | • Immunosuppressive medication<br>• Surgery to relieve pressure behind the eyes |

* Based on information supplied by the Thyroid Eye Disease Association; † If you experience this or any other eyesight changes, you need to inform the vehicle licensing authorities

## Reducing Swelling and Quelling Inflammation

If the disease is active and severe – the symptoms of inflammation are worsening – you will usually need an anti-inflammatory medication like corticosteroids, or simply 'steroids'. Another effective treatment for swelling used by some consultants is orbital radiotherapy, which involves beaming low-dose radiation at the eyes. It is important that this therapy be given by an experienced practitioner to minimise side-effects and avoid potential damage to the retina.

Corticosteroids are drugs that suppress swelling and inflammation, and lower autoimmune activity. They are also hormones derived from cortisol, a hormone produced by the adrenal glands. These hormones have many vital functions, such as helping to regulate water and salt balance, and the metabolism of protein, fat and carbohydrates. During times of stress, the adrenal glands produce extra cortisol to help the body deal with stress.

As a treatment, steroids are effective, but may come with unwelcome side-effects, including the development of a round ('moon') face, difficulty sleeping or sleep disturbance, increased appetite and weight gain, acne, mood changes, an increased risk of infections and thinning of the bones. If you have diabetes, steroids may worsen your condition by increasing blood sugar levels. For these reasons, it is extremely important to follow your doctor's instructions to the letter and not make any improvised changes in the way you take your medication.

Some specialists in thyroid eye disease prescribe azathioprine alongside steroids, a drug traditionally used to prevent rejection of transplanted organs. Azathioprine helps to tame the immune system and control autoimmune activity but, as it is a powerful immunosuppressant, it also can cause side-effects. These include heartburn, nausea and vomiting, a tendency to bruise or bleed more easily than normal, tiredness, loss of appetite, fever and chills. Again, it is very important to be closely supervised while taking this drug, with blood tests every few weeks. The objective is to quell immune activity without the need for high-dose steroid. Azathioprine is used in combination with radiotherapy in the hope that, by treating the disease early, severe complications and the need for surgery may be avoided.

## Treatment Controversies

There is still much disagreement over the most effective treatment for thyroid eye disease. One avidly disputed topic is the effect of radioiodine therapy – used to treat hyperthyroidism (*see page 109*) – on the eyes. Some doctors argue that radioiodine may bring on or aggravate thyroid eye disease by causing the release of antigens or substances the body perceives as 'foreign' into the bloodstream. Others are sceptical of this idea and yet others even claim that treating the thyroid with radioiodine could actually improve the eyes.

Important new data became available in 1998 when an influential study by a group of Italian researchers appeared in the *New England Journal of Medicine*. This report concluded that radioiodine treatment for Graves' disease could indeed provoke the appearance of eye problems or exacerbate existing ones, although the effect usually only lasted for two or three months.

Since then, the consensus appears to be that radioiodine should be avoided by those with severe or active thyroid eye disease, who should be treated with antithyroid drugs instead. For those with milder eye disease, radioiodine may be an appropriate treatment, with a three-month course of steroids starting on the day the radioiodine therapy begins.

## Can Surgery Help?

Once thyroid eye disease has entered the inactive phase – the orbital cells are no longer inflamed – anti-inflammatory medication is useless with no inflammation to quell. At this point, corrective surgery may be advised to help alleviate problems and improve the appearance of your eyes.

The most common surgical procedure is an operation called orbital decompression, designed to relieve tissue pressure in the orbits by enlarging them to allow the swollen tissue to expand. The operation may be advised to alleviate pressure on the optic nerve or severe exposure of the corneas, to protect them from damage and infections as a result of such exposure. It can also help improve the appearance of the eyes. There are various highly specialized ways to reach the orbits, but the most common is to make a small cut and enter from the side of the eyes. A more recent technique is to enter via the nostrils to avoid visible scarring. The operation should be performed by a specialist who is experienced in orbital surgery.

Squint surgery can reposition the muscles that control eye movements to improve alignment and reduce double vision. It is normally done under local anaesthesia and may require a couple of operations. Often, a stitch or two is left in place so that the surgeon can make a final correction.

Corrective eyelid surgery can be done to the upper or lower lids to remove scar tissue and excessive fatty tissue, and restore the eyelids to a more normal position.

If you decide to go ahead with any of these types of surgery, make sure that the surgeon is experienced in that particular procedure. It is also important to be realistic and acknowledge that surgery cannot usually restore your appearance to what it was before the disease.

## Living With Thyroid Eye Disease

As hard as it may be to accept, it may be wise to prepare yourself for a long, slow haul before your eyes settle down. Some doctors try to soften the blow by being vague about how long it can take for thyroid eye disease to be resolved but, as Lucy points out, this may be misplaced kindness:

*I wish my consultant had told me how long it
would take. I didn't realize how long it would be,
so I kept hoping. I desperately wanted them to be
better for my wedding and, when it became
obvious that they wouldn't be, I felt bitterly upset
and disappointed. I feel more honesty on his
behalf would have made it easier to cope.*

In the end, she found several ways to help herself:

*I use a blue gel eyemask, which you stick in the
fridge, which is really soothing. I also wear dark
glasses, which cuts the glare of the fluorescent
lights at work and also makes me feel less self-
conscious. People have got used to me wearing
them and I feel much better. I don't use much
make-up as a rule but, when I got married, I went
to a make-up artist who did wonderful things
with my eyes. She used a dark eyeshadow, which
she shaded to a darker brown towards the outside,
and then used bright green on the corners. It
looked really effective.*

Another woman with thyroid eye disease said:

*You've got to learn to live with it. If a child says
something, I'll usually ignore it, but if the parent
says something, I'll usually say, 'Would you like to
see?' You do still get down days, but you can't let
it stop your life.*

You, too, can find ways to help yourself physically and emotionally. Bear in mind that, although thyroid eye disease can
make life difficult, there is no reason why you shouldn't still be
able to make positive choices about your life and the way you
choose to lead it.

# Watchpoints

- **Make the most of yourself.** Experiment with make-up. If your eyes can tolerate it, use a dark eyeshadow, which helps to make them recede. If make-up aggravates your eye symptoms, try a bright lipstick, grab some gorgeous hair accessories, buy yourself a pretty scarf or top, or wear eye-catching jewellery to draw attention away from your eyes.
- **Master the art of disguise.** Sunglasses always look glamorous; even if you have to wear them, no one else needs to know. Make sure you try on lots of pairs before buying, and find a style that flatters your face. Wear a broad-brimmed hat to help hide your eyes.
- **Take care of your body.** No matter what the state of your eyes, you can feel instantly better about yourself if you look after yourself physically. Caring for your body can help improve your self-esteem. Nurture yourself with healthy, nutritious food, don't smoke, care for your skin and make time to exercise. You may have problems with your eyes, but that needn't stop you from having a great pair of legs or a flat stomach! Give yourself time to rest and pamper yourself by going for the occasional beauty treatment or massage.
- **Listen to your body.** In many autoimmune conditions, flare-ups are likely to occur when you are under stress. Learn to recognize the signs of stress and take steps to avoid or minimize them (*see Chapter 6*). Fatigue and other symptoms can make everything an effort. Learn to pace yourself, and make time for rest and relaxation.
- **Accentuate the positive.** Remember that however poorly you feel, the disease will eventually burn itself out and treatments are effective. Studies show that feelings of helplessness weaken the immune system, so staying optimistic may even help to slow the disease process.
- **Ask for help.** Having friends and family to talk to when you are undergoing the discomfort and physical changes

of thyroid eye disease can help you feel less isolated and afraid. Studies show that having support can lower stress. Don't be afraid to ask those around you for help and support. Showing your friends and relatives this chapter and/or literature produced by a thyroid support group can help them to understand your problems and how they may support you effectively.

- **Seek support.** Groups such as the Thyroid Eye Disease Association (*see page 234*) can be a tremendous help. Their members know what you are going through, and you can share ideas and experiences, and learn more about the disease and how others facing the same problems manage them. In some cases, counselling or psychotherapy may help.

- **Don't stop living.** People with thyroid eye disease sometimes hide away because they are afraid of the reaction they will get from others, but this only increases feelings of depression and isolation. You are still you no matter what you look like. Experiment with ways to disguise the perceived flaws in your appearance, and work out a few ripostes to adverse comments from other people – and then get out there with your head held high.

For more self-help tips, see Chapter 6.

CHAPTER NINE

# Thyroid Problems and Reproduction

In Ancient times, neck enlargement was taken as a sign of pregnancy. Today, it is well established that the thyroid gland has major effects on the reproductive cycle throughout life. This chapter looks at how thyroid problems may affect you during your reproductive years.

## The Thyroid and Menstruation

The German physician Carl von Basedow, who vies with Graves for the distinction of being the first to identify auto-immune hyperthyroidism, noticed that the condition could affect menstrual periods in 1840. Since then, the connection between thyroid problems and menstrual disorders has been further confirmed. An early diagnosis means that menstruation may be disrupted less often. In a Greek study reported in *Fertility and Sterility* in 2000, only a fifth of women with hyperthyroidism had menstrual disturbances compared with over half of those in older studies. Slightly more (23.4 per cent) women with hypothyroidism experienced menstrual distur-bances, but this proportion is still substantially less than in previous studies. However, for some women, thyroid prob-lems still play havoc with periods; for example, heavy periods may be linked to hypothyroidism.

For June, a lawyer and lecturer, menstrual problems were among the first clues that she was hypothyroid:

> *My periods were extremely erratic and, when they did come, they were very heavy. I also had bad PMS. Eventually, my GP referred me to a gynaecological clinic, where I was diagnosed as having polycystic ovary syndrome [PCOS]. I was never given a blood test for this; the diagnosis was purely on the basis of my symptoms – period irregularities, weight gain and facial hair. It wasn't until many years later that the doctor did a test and said that I didn't actually have PCOS at all. The cause of my symptoms was my thyroid. Since being treated, my periods have become regular, although they are still heavy and last for a full seven days.*

Irregular periods, as June discovered, are one of the main symptoms associated with thyroid problems and, in cases of severe hypothyroidism, there may be ovulation failure – known medically as anovulation – leading to infertility. A change in bleeding pattern is a common observation, with periods becoming light and scanty (oligomenorrhoea) or heavy (menorrhagia), especially in those who are hyperthyroid.

As June further observes, premenstrual syndrome (PMS) also appears to be more common in women with thyroid problems. PMS itself is a difficult condition to define and is not well understood. Its symptoms – physical, mental and emotional – are numerous and confusingly similar to those of thyroid problems, including an altered interest in sex, anger and aggression, anxiety, increased appetite, mood swings and insomnia. Indeed, it has been suggested that PMS is not one but many different conditions, each with a different underlying cause, one of which may be a malfunction of the hypothalamic–pituitary–thyroid interconnection.

## The Thyroid and Fertility

Difficulty in becoming pregnant is sometimes the first hint of thyroid problems, especially among those with autoimmune problems causing an underactive thyroid. Hidden or 'silent' thyroid problems may also be a factor in endometriosis and tubal blockage, both of which are due to inflammation. Thyroid problems are also more common in women with polycystic ovary syndrome (PCOS), itself a cause of reduced fertility, as well as in those with infertility for which no apparent reason can be found.

Although tests of thyroid function are carried out as part of the investigation of infertility, such tests may not be among the first to be performed. The result is that many women have to endure lengthy and expensive investigations that they might have avoided had they undergone a simple thyroid function test earlier in the proceedings.

## The Thyroid in Pregnancy and After Giving Birth

Thyroid problems often emerge during pregnancy or after giving birth, although diagnosis and treatment may be complicated by the normal changes that occur in the thyroid gland during pregnancy. Such changes can improve or worsen any existing thyroid disease. Pregnancy can also be a factor in triggering Graves' disease or, less often, Hashimoto's thyroiditis. Temporary or permanent problems of thyroid functioning often develop after birth, in particular, postpartum thyroiditis (PPT).

## Table 9.1
## Common thyroid problems during pregnancy and after giving birth

| During pregnancy | After giving birth |
|---|---|
| Autoimmune hyperthyroidism (Graves' disease) | PPT leading to an overactive thyroid (postpartum thyrotoxicosis) |
| Autoimmune hypothyroidism (Hashimoto's disease) | PPT leading to an underactive thyroid (postpartum hypothyroidism) |
| Underactive thyroid after thyroid surgery | |
| Mild (subclinical) thyroid problems | |

A tremendous amount has been learned about the effects of thyroid disease on pregnancy and the postnatal period since the first edition of this book. In particular, there is considerably more knowledge of the effects of pregnancy on the immune system and the role of the pregnancy hormone human chorionic gonadotrophin (hCG), produced by the pituitary.

## Pregnancy and the Immune System

Doctors have long recognized that becoming pregnant is an immune event. It is well known, for example, that autoimmune conditions such as rheumatoid arthritis often improve during pregnancy and flare up with renewed vigour after giving birth. Just how far-reaching the effects are has not been fully understood until fairly recently.

Pregnancy involves a delicate immunological balancing act. The unborn fetus has inherited half of its cells from your partner so it is, to all intents and purposes, a foreign invader that by all the normal laws of biology should be rejected by your immune system. According to Dr John Lazarus, pregnancy is akin to a blood transfusion in that the body 'is exposed to a large amount of intravenous antigens – substances that can trigger an immune response – that are partially self and partially foreign.'

The fact that the body doesn't mount a full-scale attack on the unborn fetus is thanks to some biochemical legerdemain resulting in decreased immune activity while allowing just enough vigilance to protect both mother and baby from any real threat to health and wellbeing. It is thought that the thyroid plays a major part in this complicated piece of trickery.

## The Secret Life of Your Hormones

During pregnancy, a complex series of biochemical changes takes place that enables the mother to cope with the physical demands of pregnancy while ensuring that her unborn baby grows and develops into a healthy individual. Among the most remarkable changes are those affecting the thyroid. From the moment of conception to birth, a chain of biochemical messages passes back and forth between the mother and baby through the placenta, the fetus' lifeline, to orchestrate changes in both maternal metabolism and development of the fetus.

As levels of the female sex hormone oestrogen rise in early pregnancy, there is a corresponding rise in levels of the thyroid-binding globulin (TBG), the protein that transports thyroid hormone around in the bloodstream. This, in turn, creates an increased demand for thyroid hormone, and a rise in the total amount of $T_4$ and $T_3$ circulating in the blood. For more thyroid hormone to be made, there is an increase in iodine metabolism and the rate at which your kidneys clear iodine from the body, as evidenced by an increase of iodine in the urine. Another key change is increased production of an enzyme (type III deiodinase) whose job it is to inactivate $T_4$ and $T_3$. This leads to an increased demand for $T_4$ and $T_3$ that boosts the levels of thyroglobulin, the protein responsible for the manufacture and storage of thyroid hormone.

## The Pregnancy Hormone

One hormone that plays a key role in these changes is human chorionic gonadotrophin (hCG) – sometimes known as the 'pregnancy hormone' – produced by the placenta. Levels of hCG rise rapidly in early pregnancy and trigger the fall in thyroid-stimulating hormone produced by the pituitary.

All this biochemical activity is visible in the slight increase in thyroid size that was one of the first 'pregnancy tests' in the Ancient world. The Egyptians would tie a reed around the neck of a young woman when she got married and, when it broke, it was a sign that she was pregnant! Ultrasound scans reveal that the thyroid increases in volume by 30 per cent between weeks 18 and 36 of pregnancy. An increasing amount of evidence is accruing to show that these changes in thyroid function may be particularly important for the healthy development of the unborn infant's brain.

Carol is convinced that her undiagnosed thyroid problems are responsible for her children's problems at school:

> *My son is dyslexic, which I understand is*
> *associated with thyroid illness – particularly where*
> *thyroid illness is undiagnosed and untreated at the*
> *time of pregnancy. My daughter has mild learning*
> *difficulties – mainly with spelling and number*
> *work – and again, I think this is linked to my*
> *illness being undiagnosed.*

## Table 9.2
## A delicate balance: thyroid changes in pregnancy

| What's happening in your body | What's happening to your thyroid |
| --- | --- |
| Increased levels of oestrogen | Increased levels of thyroid-binding globulin, which carries thyroid hormone around in the bloodstream |
| Increased levels of thyroid-binding globulin | Increased demand for $T_4$ and $T_3$ leading to an increase in total $T_4$ and $T_3$ |
| Increased levels of the 'pregnancy hormone' human chorionic gonadotrophin | Lower levels of thyroid-stimulating hormone, produced by the pituitary gland, and higher levels of free $T_4$ |
| Increased clearance of iodine by kidneys | Increased need for iodine in the diet and increased excretion of iodine in the urine |
| Increased levels of enzyme that deactivates thyroid hormone | Increased breakdown of thyroid hormone and increased demand for hormone production |
| Increased demand for thyroid hormone | Increased blood levels of thyroglobulin, the protein involved in thyroid-hormone synthesis; increased volume of thyroid gland, causing a swollen neck |

Based on 'Thyroid function during pregnancy', Clin Chem, 1999; 45: 2250–8

The relationship between hCG and the thyroid has come under scrutiny in recent years on account of hCG's remarkable similarity to thyroid-stimulating hormone (TSH). This similarity enables hCG to lock onto the TSH receptor – the 'keyhole' that allows TSH access to the thyroid. In some women, this can cause excess stimulation of the thyroid, resulting in 'gestational thyrotoxicosis' (pregnancy hyperthyroidism or hyperemesis gravidarum), a condition which causes severe vomiting. If you develop this condition, the symptoms of overactivity usually settle spontaneously as vomiting eases and antithyroid drugs are not required. The condition is rare – only some five out of every 1000 pregnant women will get it – and although potentially serious because of the loss of nutrients, it usually clears up during the second three months of pregnancy and always after the baby has been born.

Concentrations of TSH are known to drop in early pregnancy, when levels of hCG are at their highest, resulting in a corresponding rise in thyroid hormone. These interactions between hCG and the thyroid are yet another fascinating example of the ways in which different biochemicals communicate. In this situation, the embryo's developing lifeline, the placenta, directs the mother's endocrine system to provide the optimal biochemical environment for her baby.

## Are Thyroid Problems Linked to Miscarriage?

Uncontrolled, both hypothyroidism and hyperthyroidism appear to be linked to an increased risk of miscarriage. An increasing number of experts also suspect that recurrent miscarriage (three or more) may be a result of failure of the above-described immunological balancing act, with the result that the mother's body rejects the fetus. The culprits are thought to be the very same autoantibodies that cause the body to turn against itself in Graves' and Hashimoto's. According to the American Association of Clinical Endocrinologists, around six out of every 100 miscarriages may be a result of autoimmune thyroid disease during pregnancy. For this reason, all women in the US are advised to have a TSH test before becoming pregnant or early in the antenatal period, although doctors in the UK don't recommend this unless a specific problem is suspected.

Nevertheless, support for the American approach may be emerging from some influential research carried out in the 1990s which found that women who persistently miscarry are significantly more likely to have antithyroid antibodies compared with women who carried their babies to term. These findings are still the subject of much debate among thyroid experts, with some doctors suggesting that thyroid autoantibodies could be used as a marker to identify those women at risk of miscarriage and that thyroid treatment could be used to prevent them.

## Are Thyroid Problems Linked to Birth Defects?

In the past, doctors have maintained that properly treated thyroid disorders do not pose any risks to the unborn baby. However, some new and rather worrying research carried out in January 2002 suggests that this might not be entirely true. This study, carried out by researchers at Johns Hopkins University in Baltimore, Maryland, and reported at a meeting of the Society for Maternal–Fetal Medicine in New Orleans, found that babies born to women with either an overactive or underactive thyroid were more likely to have heart, brain and kidney defects as well as other, less serious abnormalities such as a cleft lip or palate, or extra fingers. In addition, babies born to hypothyroid mothers had an increased risk of heart problems even if their mothers were being treated. Dr David Nagey, the head of the Johns Hopkins research team, has called for thyroid function tests to be part of the battery of routine examinations carried out in early pregnancy. However, British experts in fetal medicine remain sceptical. But, if these findings do prove to have foundation, it could lead to the introduction of thyroid screening in early pregnancy, perhaps followed by a fetal echocardiogram (heart function test) to be carried out during week 20 of pregnancy, if hypothyroidism is detected.

## Your Baby's Thyroid

The unborn baby's thyroid gland starts to manufacture its own hormones after 10–12 weeks in the womb. Before this, your baby is totally dependent on you to supply his needs via the placenta. It's also thought that the baby may be able to absorb thyroid hormones from the liquid (amniotic fluid) in which he floats. Levels of $T_4$ thyroid hormone are low until halfway through the pregnancy and steadily rise as the pregnancy continues, while levels of the active $T_3$ hormone are high at mid-pregnancy and fall towards birth. There's increasing evidence to suggest that these thyroid hormone levels are linked to maturation of the baby's brain and nervous system.

### Stay Warm, Baby

After the baby is born, levels of thyroid-stimulating hormone (TSH) produced by his own pituitary rise abruptly as the baby starts its independent life, but fall to normal levels after a few days. Full-term babies have a good padding of fat under their skin to help them keep warm. In the first few weeks after birth, the rate at which $T_4$ is converted into active $T_3$ within this fatty tissue, it is suggested, is a crucial factor in the newborn's temperature-control mechanism.

## Hypothyroidism in Pregnancy

Hypothyroidism affects between three and seven women in every 1000 pregnancies. The most common causes are Hashimoto's thyroiditis, thyroid surgery or radiation treatment for an overactive thyroid. The good news is that expectant mothers with autoimmune disorders – including Hashimoto's – often find that their condition improves during the latter part of pregnancy although, sadly, there's often a worsening once the baby is born.

In the past, when thyroid problems weren't so well managed, hypothyroidism was associated with a higher risk of stillbirth, miscarriage and complications such as preeclampsia (high blood pressure in pregnancy), bleeding, low birth weight and birth defects.

## Planning for Pregnancy

If you are hypothyroid, your dose of $T_4$ will need to be increased during pregnancy by 25–50 microgrammes on average. This will need to be done from the time of conception to tide your baby over until he can produce his own thyroid hormone – from about 12 weeks. Until then, your baby is entirely dependent upon you. Unfortunately, not all GPs are aware of this, so if you're planning a pregnancy, make sure to arrange with your doctor for your treatment to be adjusted. Rest assured that taking thyroid medication in pregnancy is perfectly safe. Indeed, it is essential to ensure your baby's healthy growth and development.

## Diagnostic Difficulties

Diagnosis of an underactive thyroid during pregnancy can be tricky as, even in a straightforward pregnancy, pregnant women may experience similar symptoms to hypothyroidism, such as extra sensitivity to cold, coarse hair, difficulty concentrating and irritability. If the doctor suspects a thyroid problem or if there is a family history of thyroid disease, he will probably suggest a TSH blood test and a check-up for the presence of the tell-tale antibodies associated with autoimmunity. It's worth bearing in mind that the interpretation of these tests may be rendered more difficult by the normal physiological changes of pregnancy.

## Treatment Matters

If you have developed hypothyroidism for the first time during pregnancy, you will be prescribed a treatment (*see Chapter 5*), and the doctor will want to keep a careful eye on you during your pregnancy and during labour to ensure that you and your baby remain fit and well.

# The Thyroid and IQ

Over the past few years, there has been a growing realization of just how important thyroid hormones are for the development of the unborn baby's brain and nervous system. It has long been known that babies born with an underactive thyroid (congenital hypothyroidism) – due to absence of a thyroid or failure of the gland to develop or function – would grow up mentally and physically retarded (cretinism). This condition can still be seen in developing parts of the world where the soil is iodine-deficient. Fortunately, it has all but disappeared in the developed countries, where all babies are given a blood test five to seven days after birth to measure thyroid function and, if necessary, given thyroid replacement therapy.

It has only been fairly recently that concerns have emerged over the effects of mildly low levels of thyroid hormone during pregnancy on the intellectual development of the unborn baby. The warning bells began when a study carried out in 1991 suggested that low levels of thyroid hormone in early pregnancy could adversely affect IQ (intelligence quotient). In 1999, a study published in the *New England Journal of Medicine* revealed that even a mild (subclinical) shortage of thyroid hormone during pregnancy could affect a child's IQ years later. Specifically, children aged seven to nine whose mothers were hypothyroid during pregnancy had IQ scores that were about seven points lower than those whose mothers had no thyroid problems. Since then, other research has shown that the children whose mothers had a particular kind of antithyroid antibody also had lower IQ scores.

These concerns have led to calls from some quarters for thyroid function tests to become a routine part of the raft of antenatal blood tests carried out in early pregnancy or even when trying to conceive. Such screening might also identify those women at risk of postnatal thyroiditis.

## Antenatal Detection?

There are powerful arguments for making thyroid screening part of the routine antenatal programme. The fact that some 2.5 per cent of women have a raised level of TSH (thyroid-stimulating hormone) during pregnancy, suggestive of a failing thyroid, is only one of them. The fact that 10 per cent of women have antithyroid antibodies is another good reason, especially when you consider that such antibodies are linked with malfunction of the thyroid and the development of post-partum thyroiditis or PPT (*see page 194*), which may be confused with postnatal depression. Even so, routine thyroid screening could raise some tricky questions – for example, how should women found to have a mild thyroid deficiency be treated and how might such treatment affect her baby? It's to be hoped that, in the next few years, studies will be carried out to provide answers to these questions.

## Hyperthyroidism in Pregnancy

Even though hyperthyroidism is more common than hypothyroidism during pregnancy, only about two in every 1000 mothers-to-be are affected. In 85–90 per cent of cases, the overactivity is a result of Graves' disease. Overactive (thyrotoxic) nodular goitres – which tend to affect older women – are, as you might expect, unusual. More rare still are conditions such as gestational thyrotoxicosis (*see pages 184–5*) and hydatidiform mole (molar pregnancy), where the placenta grows rapidly and abnormally to resemble a bunch of grapes, as the cause of hyperthyroidism in pregnancy.

## Getting Ready for Pregnancy

Untreated hyperthyroidism can lead to a greater risk of miscarriage, premature birth, low birth weight, stillbirth and birth defects. It is also associated with a number of pregnancy complications, including accidental bleeding caused by the placenta starting to separate (placental abruption), early labour and an increased risk of preeclampsia, the hypertensive disease of pregnancy. For these reasons, it's important that your condition is properly controlled during pregnancy.

If you have Graves' disease, it would be a good idea to see the doctor in charge of your treatment before you conceive to get advice on how your condition may best be managed during pregnancy and after giving birth.

Arrangements should be made for your thyroid function to be tested more often (every four to six weeks) and for any adjustments to be made to your medication. Many doctors prefer to use propylthiouracil (PTU) rather than other antithyroid medications during pregnancy, which have occasionally been associated with rare complications. As Graves' disease often clears up by itself during the latter part of pregnancy as a result of your immune system being suppressed, it is possible that you will not need any further treatment until after the baby is born.

The doctor should tell you of any potential effects on your baby and discuss breastfeeding (*see page 202*). Approximately one in 100 babies exposed to antithyroid drugs while in the womb develop a temporarily underactive thyroid after birth, a condition known as transient neonatal hypothyroidism (*see page 200*).

There is also a small risk that your baby may develop temporary overactivity of the thyroid or transient neonatal hyperthyroidism. This can happen even if you have previously been treated for your overactive thyroid and now have normal thyroid function or are hypothyroid. The doctor should arrange for you to have blood tests to check your and your baby's thyroid function before planning the most appropriate management programme.

Although you may understandably be cautious about taking medications during pregnancy, it is reassuring to know that antithyroid treatments have been used for many years with no long-lasting effects on the baby either before or after birth, except for the temporary effects mentioned above. It's also worth bearing in mind that if hyperthyroidism is not adequately treated, this in itself can adversely affect the baby.

## Hyperthyroidism Developing During Pregnancy

Again, it can be difficult to diagnose an overactive thyroid during pregnancy because so many symptoms, such as heat intolerance, rapid heartbeat, nervousness and an enlarged thyroid, are seen during normal pregnancy. The doctor should check for poor weight gain, goitre, lid lag (*see page 164*), muscle weakness and rapid heart rate, and perform a thyroid function test to determine how well your thyroid is working and whether you have any thyroid autoantibodies (*see Chapter 4*). The presence of certain antibodies may suggest that the baby is likely to develop a transiently overactive thyroid after birth. It may also be a warning that you are likely to develop postpartum thyroid disease (*see page 194*).

As far as treatment is concerned, most doctors will opt for medication rather than radioactive iodine or surgery. Getting the dosage right is vital as antithyroid drugs can cross the placenta and, if the amounts taken are too large, the unborn baby may develop a goitre, which could cause problems during labour and delivery. Most doctors are also cautious about using block-and-replace therapy during pregnancy for the same reason.

## When Surgery is Needed

Clearly, most surgeons avoid performing an operation during pregnancy. However, if the hyperthyroidism is severe and doesn't respond to treatment, or the treatment is causing severe side-effects, a partial thyroidectomy may be suggested. You may also need surgery if you have a very large goitre. If surgery is necessary, it will usually be performed during the second trimester (months three to six) of pregnancy. Nevertheless, this is definitely a last resort.

## If You Discover a Lump

Around one in 10 women develop thyroid nodules in pregnancy. If you discover a lump during pregnancy, it will need to be investigated by fine-needle aspiration biopsy (*see Chapter 3*) to establish whether it is benign or malignant (cancerous). If it turns out to be cancerous, the surgeon may decide to operate during the second trimester or, in some cases, it may be decided to leave surgery until after the baby is born. Because thyroid cancer tends to develop slowly (*see Chapter 3*), its impact on pregnancy seems to be minimal and the risk of the cancer spreading or recurring is the same as in a woman who is not pregnant. However, it is worth discussing the pros and cons of operating during pregnancy with your surgeon so that you can feel sure that you've come to the decision that is right for you.

## Pregnancy Watchpoints

- If you have a preexisting thyroid problem, make an appointment to see your doctor before you conceive to discuss how your condition may change during pregnancy and what steps can be taken to manage it.
- Make sure to attend all your antenatal appointments and that your thyroid problems are included in your notes. Your GP, midwife and the obstetrician overseeing your care should know that you have a thyroid disorder.
- Although you don't need to eat for two in terms of quantity, your demand for nutrients, including iodine, is increased during pregnancy. It is particularly important to

make sure you eat a healthy, well-balanced diet, with plenty of fresh fruit and vegetables.

- Keeping a balance between activity and rest is important. Fatigue can be a real problem as the natural fatigue of pregnancy may be exacerbated by thyroid problems. Make sure you make time for light activity and relaxation. Don't feel you are letting your side down if you take time out to rest.

- If you develop severe vomiting to the extent that you are losing weight and/or can't keep anything down, contact your doctor. You may have gestational thyrotoxicosis and will need to be admitted to hospital to be rehydrated.

- If, like many women, you are keeping a pregnancy diary, use this to keep a record of your physical health. This can help you identify any change in symptoms that may need medical attention.

- If you are hypothyroid, your dosage is likely to need adjustment during pregnancy to meet your body's increased need for thyroid hormone. While pregnant, you should have your blood tested every four to six weeks to check your thyroid function.

- If you are taking antithyroid medication to keep high levels of thyroid hormone under control, you may need to continue this at a lower dose throughout pregnancy and during labour.

- When you go into labour, make sure the midwives attending you know that you have thyroid disease.

- After delivery, make sure the paediatrician is aware you have a thyroid problem in case your newborn baby develops complications. (Your thyroid specialist should have informed your paediatrician.)

## Postpartum Thyroid Dysfunction

The most common problem after pregnancy is a temporary disturbance of thyroid activity – either overactivity or underactivity

– that may arise at any time during the first year after delivery. The most common cause of such thyroid dysfunction is post-partum thyroiditis (PPT), an inflammation of the thyroid that typically causes symptoms of overactivity, followed by a longer-lasting underactive phase. Like so many thyroid problems, PPT is thought to be autoimmune in origin.

PPT was first reported in the *British Medical Journal* in 1948 by a GP in New Zealand, although little attention was paid until a study of six Japanese women was published in 1976. This was followed by further research from Japan, Canada, Sweden, the USA and Wales. Today, the condition is well recognized.

However, despite the increased knowledge, it is still not known exactly how many women are affected; estimates vary wildly, from 2–21 per cent, probably because different studies have used different diagnostic criteria. What is known is that you are three times more likely to develop PPT if you have type 1 (insulin-dependent) diabetes or have had a previous bout of PPT.

## What Are the Symptoms?

Typically, PPT strikes within six months of giving birth with an episode of thyroid overactivity followed by a period of thyroid underactivity, usually around four to eight months after your baby's birth. Very occasionally, it happens the other way round. However, some women only experience the overactive phase while others only develop symptoms of underactivity. Whatever the course of events, PPT usually disappears and thyroid function returns to normal.

## A Difficult Diagnosis

Despite being such a common condition, PPT all too often goes unrecognized because its symptoms of fatigue, lack of energy, irritability, dry hair, hair loss, dry skin, difficulty remembering and concentrating, and depression so often become confused with the physical and mental changes experienced by so many women after birth and/or with postnatal depression.

Sarah, a social worker, describes how she developed the symptoms of an overactive thyroid after the birth of her second child:

> *I suspect that my problems actually started when I was expecting my baby. I put on hardly any weight during pregnancy and that disappeared as soon as I had the baby. By the time I had my six-week check-up, I was feeling very unwell and really not myself. I experienced terrible mood swings. Normally I'm quite jolly and even-tempered, but I was anxious and short-tempered. I would snap people's heads off for the slightest reason. I was constantly rowing with my husband, which again is not like me at all. At other times I would get depressed. I was always tired. And I became very thin, around 6 stone 8 pounds. I suffered from backache. As time went on, my muscles became weaker and weaker. If I knelt on the floor, I had to have someone help me back up again. I had terrible palpitations, and I was really worried that I was going to have a heart attack. In the end, it was actually a reflexologist who suggested that I should have my thyroid checked: she could feel enormous crystals in the part of my foot that relates to the thyroid. When I went to the doctor, he diagnosed an overactive thyroid.*

Maggie, who developed an underactive thyroid after her first baby was born, was told by her doctor that she was suffering from postnatal depression.

> *Although he's an excellent doctor, I felt he didn't really take my physical complaints seriously. At the*

*hospital, it was clear they thought I was stark*
*raving mad, and I was beginning to feel that*
*I was as well.*

If you don't feel your doctor is taking you seriously, it's worth persevering and perhaps asking for a second opinion. The illness, in its overactive phase, can be confused with Graves' disease. However, unlike Graves', PPT goes away and no treatment is usually needed unless symptoms are severe. In this case, a course of beta-blockers may be useful to control the rapid heartbeat. On the whole, antithyroid medication is not useful for PPT as, unlike other forms of hyperthyroidism, the symptoms of overactivity are not a result of too much thyroid hormone due to faulty thyroid function, but the result of inflammation of the thyroid caused by thyroid antibodies.

Similarly, as the hypothyroid phase of PPT tends to disappear on its own, treatment is not always needed unless symptoms are particularly severe.

Of course, if you have Graves' disease, it needs proper treatment with antithyroid drugs.

## Keeping an Eye on Symptoms

Although PPT usually disappears of its own accord over time, there is a 70 per cent risk of it reappearing after any future births. Another concern is that, although the hyperthyroid phase always resolves itself, in some 20–30 per cent of women, the symptoms of underactivity persist and may even progress to full-blown, permanent hypothyroidism. Research shows that around half of all women who develop PPT will be permanently hypothyroid within seven to nine years.

For this reason, it is wise to keep an eye on your symptoms and to have your thyroid function monitored at yearly intervals, or sooner, if you become pregnant again or develop any symptoms of overt thyroid disease. The fact that you had PPT should be mentioned on your medical notes. But in case it's overlooked, if you do become pregnant again, you should mention it to your doctor and the hospital on your first book-ing-in visit so that you receive any treatment you need as early as possible.

## The Depression Connection

Postnatal depression strikes one in 10 women, although some experts believe the true incidence is even higher. In the past, many reasons have been put forward to explain how it develops. These include social and psychological factors, such as relationship difficulties, housing problems and a previous history of emotional upsets. The latest research, however, suggests a physical cause in some cases – namely, the thyroid.

Dr John Lazarus, of the University of Wales College of Medicine, discovered that even healthy women without overt thyroid problems who have antithyroid antibodies have a higher risk of developing mild-to-moderate depression after delivery. In a later publication, Dr Lazarus speculated that thyroid autoantibodies could in some way modulate the activity of neurotransmitters, the chemical messengers produced by the brain and nervous system to pass along messages from the brain to other parts of the body. Clearly, more research needs to be done to see if these connections are indeed true and, if so, what part they may play in the development of postnatal depression.

### Dealing With Depression

At a time when everyone expects you to be brimming with happiness, postnatal depression can be a cruel blow. In severe cases, antidepressants may help and, if thyroid problems are a factor, thyroxine ($T_4$). In milder cases, counselling may be more appropriate. There are also many ways you can help yourself:

- Allow yourself to cry or get really angry – calm sometimes follows the storm
- Get physical – do some exercise, such as walking, running, dancing, swimming or aerobics
- Visit a friend – make it someone who cheers you up
- Go out – don't sit at home moping
- Visualize waves of energy flowing through your body and outwards to your baby and those you love
- Keep a selection of songs on hand that you find cheering or uplifting, and play or sing them when you need a lift
- If the depression persists, see your doctor, counsellor or therapist as drugs or psychotherapy may help.

## Your Baby's Thyroid

As we've seen, baby and mother are closely interlinked during pregnancy. If you have a thyroid problem, this means that substances from your bloodstream, such as medications and sometimes thyroid autoantibodies, can cross the placenta and affect your baby.

### Hyperthyroid Problems Affecting Your Baby

If you have Graves' disease, your baby may occasionally develop hyperthyroidism in the uterus and/or after birth, a result of certain autoimmune antibodies that turn on thyroid activity crossing the placenta. This situation is estimated to affect from six in 1000 to around nine in 100 mothers with Graves' disease.

A blood test for these antibodies during pregnancy can identify if your baby is at risk. Another clue may be that your unborn baby's heartbeat is particularly fast (over 160 beats per minute) after 22 weeks of pregnancy. Your baby may be particularly at risk if you have previously had a stillbirth. Untreated, the condition can lead to a number of complications during pregnancy and after birth, including poor growth, small head size and fetal distress during labour. If you think you're at risk, your baby's heartbeat should be checked regularly, and you should be checked for autoantibodies. Treatment consists of antithyroid medication for you that is sufficient to maintain your baby's heart rate at around 140 beats per minute.

If your thyroid was overactive during your pregnancy or, indeed, if you have ever been hyperthyroid, your baby will be at risk of neonatal hyperthyroidism, caused by autoantibodies crossing the placenta. Mild cases usually require no treatment as the symptoms subside as the autoantibodies pass out of the baby's system (usually within five months) and thyroid function becomes normal. In more severe cases, the baby may be quite unwell, with a rapid heartbeat, irritability and difficulty in settling, jaundice and feeding problems, and be slow to put on weight. In such cases, the doctor will prescribe antithyroid medication, beta-blockers to regulate heartbeat and perhaps sedatives to help calm the irritability.

## Hypothyroid Problems Affecting Your Baby

Antithyroid medications and sometimes blocking antibodies, which switch off the thyroid gland, may cross the placenta, causing your unborn baby to develop an underactive thyroid. This condition is more difficult to treat in the unborn baby as only small amounts of thyroid hormones are able to cross the placenta. Therefore, the doctor may prescribe a drug that mimics the activity of thyroid hormone and is able to cross the placenta more easily. Again, however, the baby is at risk of continuing underactivity after birth (neonatal hypothyroidism), although the condition resolves itself as the antibodies or drugs pass out of the baby's system over the first few months of life. Such transient (temporary) neonatal hypothyroidism can also occur in areas of the world where there is not enough iodine in the soil.

More seriously, in around one in 4000 births, the infant is born with an underactive thyroid regardless of any thyroid problems the mother may have. A newborn who is born with a shortage of thyroid hormones has what is called 'congenital hypothyroidism'. Left untreated, it can lead to brain damage, and physical and mental development problems.

Congenital hypothyroidism is one of the most common conditions of a family of so-called metabolic disorders, and is usually due to the complete absence of a thyroid gland, or failure of the thyroid gland to develop or function properly. This can arise as a result of a faulty immune system, inborn errors of thyroid function, faults in the way the body processes iodine or environmental factors, such as the mother taking certain iodine-containing drugs during the pregnancy.

Treatment with thyroid hormone enables the baby to develop completely normally. Fortunately, this problem is virtually unknown in the developed countries since all newborns undergo a heel-prick test to measure thyroid function. If the baby is found to have a disorder, thyroid replacement therapy will prevent any further problems.

## Table 9.3 Thyroid disease and your baby

| Baby's problem | Mother's problem | Cause |
|---|---|---|
| Fetal hyperthyroidism | Graves' disease | Autoimmune antibodies passing through the placenta |
| Fetal hypothyroidism | Hashimoto's disease | Autoimmune antibodies crossing the placenta, or excess antithyroid pills or drugs containing iodine |
| Fetal goitre autoimmune | Goitre in mother or Hashimoto's disease | Iodine-containing drugs, antithyroid pills, antibodies |
| Neonatal hyperthyroidism | Graves' disease or Hashimoto's disease | Autoimmune antibodies crossing the placenta |
| Neonatal hypothyroidism | Endemic goitre | Severe iodine shortage, thyroid growth-stimulating antibodies crossing the placenta |
| | Hashimoto's disease | Autoimmune antibodies crossing the placenta |
| Neonatal pseudohypothyroidism | Drugs taken by mother | Iodine-containing or antithyroid drugs; high thyroid-stimulating hormone levels caused by thyroid-stimulating hormone antibodies in the blood |

## Will I Be Able to Breastfeed?

If you want to breastfeed your baby, there's no reason why you shouldn't, although there are a few special considerations you may want to take into account. If you are being treated for hypothyroidism, there should be no problems in breastfeeding as long as your blood hormone levels stay within normal ranges.

The question of whether to breastfeed if you are being treated for hyperthyroidism is more complicated. Many doctors argue that it's not a problem, although most prefer to prescribe propylthiouracil (PTU) for a hyperthyroid mother-to-be rather than any other antithyroid compound. However, as small amounts of PTU do pass into the breastmilk, your baby's thyroid function will need to be closely monitored to ensure that the gland remains healthy. The drug methimazole, used to treat hyperthyroidism in North America, is not recommended for breastfeeding mothers.

All forms of iodine pass into the breastmilk and can cause the infant to develop a goitre. If you need a scan to diagnose an overactive thyroid, the doctor will usually suggest a scan using the metallic element technetium, rather than radioactive iodine, as it passes out of the system quicker, thus allowing you to start breastfeeding again within 30 hours.

You should also be aware that both hyperthyroidism and hypothyroidism can affect milk production. If you need help with breastfeeding, your midwife and/or health visitor can give you any support you need. You may also wish to contact a breastfeeding support group.

# CHAPTER TEN

# The Menopause and Beyond

The time of life when thyroid problems are often relegated to the back-burner is during the menopause and beyond ironically because they become increasingly common as we get older. 'Silent' (mild or subclinical) thyroid problems affect women particularly in their 40s and 50s, when they can complicate or even be disguised as menopausal problems such as depression. An estimated one in 10 women have abnormal levels of thyroid-stimulating hormone (TSH) by age 50, a hint that the thyroid may be beginning to fail; by age 60, this has risen to 17 per cent. Furthermore, over one in 10 elderly women has thyroid autoimmune antibodies, another clue pointing to potential, silent thyroid problems.

## Defining the Terms

Strictly speaking, the term menopause refers to the cessation of menstruation, although most of us use it rather loosely to describe the years approaching the menopause. During this time, levels of oestrogen and, to a lesser extent, progesterone – hormones responsible for preparing your body for a potential pregnancy – begin to dwindle, at first gradually and later steeply. Periods become irregular and the bleeding pattern may change whereas other symptoms of a lack of oestrogen, such as hot flushes and vaginal dryness, may start to become problematical.

Menopause proper – that is after you've had your last period – usually happens between ages 50 and 55. With better diets and improved healthcare, most of us can look forward to living a third of our lives after the menopause (post-menopause), a time of life that, provided we remain healthy, can be a period of increased zest and freedom. However, it can also be a time when women are particularly vulnerable to developing serious health problems caused by the lack of oestrogen – namely, brittle-bone disease or osteoporosis, caused by the loss of bone that happens as women get older, and heart disease, which is linked to higher levels of blood fats such as cholesterol. These two problems are especially relevant to anyone with a thyroid problem because faulty thyroid function is also associated with an increased risk of these problems.

## Hormones at the Menopause

The period around the menopause is a time of hormonal upheaval similar to adolescence and pregnancy, characterized by alterations in levels of virtually all of the body's hormones, but especially oestrogen. The close links between the different players in the hormonal orchestra means that thyroid problems can both affect and be affected by these other endocrine system changes.

Just as the thyroid is controlled by chemical messengers secreted by the hypothalamus in the brain and the pituitary gland, so is the control of the ovaries also under the direction of hormones produced by these two organs – specifically, follicle-stimulating hormone (FSH) produced by the pituitary to stimulate the egg to grow within its follicle (egg sac), and luteinizing hormone (LH), which causes the egg to burst from its sac at ovulation. When the body begins to run out of eggs and oestrogen levels begin to fall, more FSH is produced in an attempt to spur the failing ovaries into action. It is this rise in FSH that is thought to be responsible for the emotional symptoms of the menopause, including mood swings, increased irritability and depression. Even in healthy women without thyroid problems, there's also a natural decrease in thyroid function.

## How Thyroid Problems Complicate the Menopause

A host of nagging health problems have been attributed to the menopause. Medically speaking, the only ones considered to be truly due to a lack of oestrogen are hot flushes and dry vagina and, years later, osteoporosis and heart disease.

One of the biggest problems for anyone with a thyroid disorder, especially those with a mild malfunction that may be difficult to detect, is distinguishing symptoms of the menopause from those of faulty thyroid function. A study carried out in 1998, for example, found that depression in women around the time of the menopause was frequently not a symptom of menopause, but of potential thyroid problems. Other symptoms too – such as fatigue, mood swings, sleep disturbances and changes in the hair, skin and nails, all commonly attributed to the menopause – may actually be signs of a failing thyroid, especially if they aren't helped by hormone replacement therapy (HRT), according to the American Association of Clinical Endocrinologists (AACE). The AACE recommends that all women over 40 should have a thyroid function test, although British experts tend to be more cautious unless there are reasons to suspect thyroid problems.

## Should You Consider HRT?

Every woman approaching the menopause has to decide for herself her position on HRT, which involves replacing the body's natural oestrogen (and, if you have a uterus, proges- terone also) to avoid short- and longer-term health problems.

Many women report a significant decrease in menopausal symptoms if they take HRT and, more important, HRT can help to prevent the accelerated bone loss associated with osteoporosis. There is also some recent research suggesting that HRT may help prevent dementia. However, it is not a simple decision and, in the past few years, some doctors have become considerably less enthusiastic about HRT as more evidence is uncovered of its potential drawbacks. It has long been known that taking HRT for five years or more increases the risk of breast cancer, and now there is also a question mark hanging over the benefits of HRT on the heart.

It has long been believed that HRT protects the heart by preventing the furring of the arteries that leads to heart disease. It was thought that, in younger women, it is partly oestrogen that helped to keep the arteries healthy by lowering levels of the 'bad' LDL-type cholesterol. But recent research has somewhat dented HRT's reputation as a cardioprotector. In particular, the results of the large American Heart and Estrogen Replacement Study (HERS) reported in 2000 have cast doubts on the ability of HRT to slow heart disease, especially in women who already had furring of the arteries. This clearly has implications for women with thyroid problems, many of whom already have high cholesterol levels and an increased risk of heart disease. More research is needed to throw further light on the exact effects of HRT on the blood vessels.

## HRT and Your Thyroid

If you do decide to use HRT, what effect is this likely to have on your thyroid? Probably very little, according to a report in the *New England Journal of Medicine* in 2001 – provided you have a healthy thyroid. Just as they do in pregnancy, the higher levels of oestrogen stimulate a rise in blood levels of the thyroid transport protein thyroxine-binding globulin (TBG); in women without thyroid problems, the body quickly adapts to the TBG increase. But if you have hypothyroidism, the increased TBG is not so quick to settle and, indeed, can cause a drop in free thyroid hormone (hormone not bound to proteins in the bloodstream) that, in turn, can create a need for higher doses of thyroxine ($T_4$). So, if you are hypothyroid, you may start to feel worse unless your dose is increased. You should undergo a thyroid function test 12 weeks after beginning to take HRT so that your dosage of $T_4$ can be adjusted if necessary.

## Thyroid Problems and Osteoporosis

With one in four women at risk of developing osteoporosis following the menopause, the effect of the thyroid on bone is of particular importance. Osteoporosis leads to an increased risk of fractures in later life and is the cause of much ill health in older women. Thyroid hormone speeds up the breakdown and natural replacement of bone that is ongoing throughout our lives. It is well known that women with an overactive thyroid are at risk of 'secondary' osteoporosis, weakness of the bones as a result of an underlying condition – in this case, the hyperthyroidism – and that restoring thyroid function to normal eliminates the risk. Unfortunately, as we've already seen, it can take some time to find the right medication, during which time, bone may continue to be lost.

According to some studies, $T_4$ can block the beneficial effects of HRT on bone density in women with mildly underactive thyroids. A small study reported in 2001 suggested that long-term (more than 10 years) $T_4$ treatment as you approach the menopause can result in osteopenia, lower bone density, by the start of the menopause. Other research from Israel found that $T_4$ could interfere with the bone-building benefits of HRT in women with mild hypothyroidism. One study reported in the journal *Clinical Endocrinology* in 1997 showed that, in women past the menopause taking $T_4$, bone turnover was closely linked to blood levels of thyroid-stimulating hormone. This study suggested that reducing the $T_4$ dose in women with low TSH levels could decrease the rate of bone turnover and increase bone density. These findings all emphasize the importance of finding the right thyroid medication and dosage as well as doing all you can to look after your bones (*see below*).

## Looking After Your Bones

- Eat a bone-friendly diet by making sure you consume lots of calcium-rich foods like milk, cheese and other milk-based dairy products, sardines and shellfish, green leafy vegetables, citrus fruits, nuts and seeds.
- Take weight-bearing exercise. Anything that 'loads' your bones, such as running, dancing or resistance work with weights, can help rebuild bone and slow bone loss.
- Watch your alcohol and caffeine intake. Both increase the loss of calcium through the urine.
- Ask your doctor or consult a nutritional practitioner about taking a calcium supplement.
- Quit smoking. Smoking is likely to lead to increased bone loss, and smokers tend to enter the menopause earlier than non-smokers. Smoking is also associated with an increased risk of autoimmune thyroid problems.
- Ask your doctor if he recommends a bone scan. This can show your bone density and whether you are at risk of osteoporosis.
- Seek advice on taking HRT. If you decide to take it, make sure the doctor who prescribes it knows that you have a thyroid problem, and bear in mind that your thyroid medication may need to be adjusted. Ideally, you should have a thyroid function test 12 weeks after starting HRT.

- If you can't or don't wish to take HRT, discuss alternatives with your doctor. There are other medications that can help prevent bone loss.
- Get your thyroid treatment right. It's important for the health of your bones that your thyroid disease is properly controlled. If you aren't happy with your treatment, don't give up. Keep going back to the doctor until your thyroid function is back to normal.
- Don't miss your thyroid check-ups. As dosages need changing with changed hormone levels and weight gain, it's important to carry on having regular thyroid function tests. It's tempting, especially if you're feeling well, to skip these – don't!

Find out more about eating for healthy bones in Chapter 6.

## Fats, Fat and Heart Disease

Given the association between low levels of thyroid hormone and weight gain, could the natural dip in thyroid hormone during the run up to the menopause be linked to the difficulty in losing weight that so many women complain of at around this time? More serious, could it contribute to the rise in unhealthy blood fats with the accompanying greater risk of heart disease? It's an appealing theory but, so far, the answers are not clear.

In a study of women participating in the Healthy Women Study, a large-scale study of heart disease risk factors at the menopause reported in the *Journal of Women's Health* in 1997, researchers from the University of Pittsburgh evaluated thyroid function, and changes in weight and levels of blood fats. The women, aged between 42 and 50, were healthy with no apparent thyroid dysfunction. The researchers concluded that any changes in thyroid function in these women were unlikely to cause weight gain or raised blood fat levels.

What the researchers did discover, however, was that women with thyroid antibodies – which may be a sign of hidden thyroid problems – had lower levels of total cholesterol and of 'bad' LDL cholesterol, associated with furring of the arteries. As with bones, this study underlines the importance of having your thyroid medication checked, any problems treated and looking after your heart.

## Looking After Your Heart

- Consume a diet rich in fruit and vegetables. These are rich in antioxidant vitamins that help prevent furring of the arteries.
- Take regular aerobic exercise. Experts advise 30 minutes of any exercise that makes you huff and puff slightly on most days of the week to maintain a healthy heart.
- Eat oily fish. Fish oils are protective, so try to include oily fish such as salmon, tuna, swordfish and herring on the menu at least three times a week.
- Get your blood pressure checked. High blood pressure is a risk factor for heart disease and stroke. It is known as a silent killer because it usually doesn't cause any symptoms. Regular blood pressure checks can help ensure that your blood pressure doesn't become dangerously high.
- Know your cholesterol. If you have a thyroid problem, realize that you are at an increased risk for heart disease and ask your doctor to check your cholesterol levels.

## Thyroid Problems and Breast Cancer

There's also a question mark over whether thyroid disease may be linked to breast cancer, another condition of particular concern for women past the menopause. Some studies suggest that women with underlying autoimmune thyroid disease have a higher risk of developing breast cancer. Reassuringly, a recent study of nearly 15,000 patients found no such association. It could be that previous associations were due to the fact that both thyroid problems and breast cancer tend to become more common in later life. However, it would still pay to be vigilant.

## A Word About Soy

One issue of particular relevance to women with thyroid problems is whether a diet high in soy can improve health at the menopause. Soy is a rich source of phytoestrogens (plant oestrogens), which many nutritionists believe can help protect against menopausal symptoms, such as hot flushes, night sweats and dry vagina, as well as breast cancer while improving cardiovascular health. The problem, if you have thyroid disease, is that soy is also a goitrogen (a goitre-producing substance) and, in some studies, has been found to reduce absorption of thyroid hormone and increase the need for $T_4$ medication.

A number of nutritionists recommend soy as a kind of 'alternative HRT' for menopausal women who can't or don't wish to take HRT. However, other doctors and nutritionists believe that anyone with hypothyroidism should steer clear of soy products. So what should you do? In her book *The Natural Health Handbook for Women* (Piatkus), UK-based nutritional practitioner Dr Marilyn Glenville advises that soy is best eaten in its traditional forms, such as miso, tofu or soya milks made from whole beans, and counsels that it's best to avoid soya bars, snacks and other overprocessed soy products. You should also aim to include other phytoestrogens in your diet, such as chickpeas, garlic, lentils, seeds and grains, and herbs such as sage, fennel and parsley.

Find out more about diet and exercise in Chapter 6.

## Thyroid Problems in Later Life

As we get older, the thyroid gland undergoes changes in structure leading to naturally lower levels of $T_3$ and slightly higher levels of TSH, the thyroid-stimulating hormone that triggers thyroid hormone production. The body also becomes less efficient at converting $T_4$ to active $T_3$. It's thought that this 'natural hypothyroidism' may be the body's way of downregulating thyroid activity to match our lower energy needs as we grow older. This appears to be supported by animal studies showing that administering extra thyroid hormone shortens the lifespan whereas inducing a state of hypothyroidism appears to lengthen it. It is thought that higher circulating levels of thyroid hormone in men compared with women may account for why women live longer.

Another intriguing finding from animal research suggests that limiting food intake, which decreases thyroid hormone production, slows down ageing and is associated with fewer age-linked degenerative illnesses. It is not known whether this would apply to humans, but there is some suggestion that it may.

Ultimately, both hypothyroidism and hyperthyroidism can lead to ill health and a poorer quality of life as we get older.

### Hypothyroidism in Later Life

An underactive thyroid becomes more common as we age, and mild hypothyroidism caused by underlying Hashimoto's thyroiditis may be a particular problem. According to the American Association of Clinical Endocrinologists, around 20 per cent of women over 60 have some form of thyroid disease. Unfortunately, it is often missed because symptoms like a slower metabolism, higher levels of blood fats, furring of the arteries, dry skin, greying hair, slower reflexes and poorer mental performance are similar to those associated with 'normal' ageing.

According to the Women's Health and Ageing Study, a large US study following the health of a group of women as they age, even just slightly lower levels of thyroid hormone may increase the risk of mental decline in the over-65s. Although the study doesn't prove that hypothyroidism affects mental sharpness as we get older, it does raise the worrying possibility that even a mild underfunction left untreated may affect mental acuity.

## Hyperthyroidism in Later Life

Thyroid overactivity also becomes more common as we age. If you are 60-plus, you are seven times more likely to develop hyperthyroidism than someone younger. Again though, symptoms may go unrecognized. In fact, some doctors believe there is a vast underdiagnosis of hyperthyroidism in the over-60s. This may be due to women in this age group being unwilling to bother the doctor with such vague symptoms, but it may also reflect the fact that doctors don't always take older women's health concerns seriously or that they tend to attribute the symptoms of hyperthyroidism to simply growing old. Whatever the reason, it is worrying because of the association of thyroid disease with other health problems.

Hyperthyroidism can be particularly difficult to detect if you are over 60 because it can be 'masked' (*see page 116*). Symptoms include:

- unexplained weight loss; as older people are prone to weight loss because of a lack of activity and/or poor appetite, this symptom may be easy to miss
- atrial fibrillation, a term used to describe an irregular, rapid heart beat that may be accompanied by shortness of breath, chest discomfort and dizziness; other symptoms

include fainting or weakness due to the heart's weakened ability to pump
- gastrointestinal problems such as constipation
- depression and/or confusion, which may be erroneously attributed to other conditions such as dementia
- bone pain and/or fractures.

These symptoms aren't always indications of an overactive thyroid. However, if you do experience any of them, it may be worthwhile making an appointment to see your GP and asking for a thyroid test.

## Treatment Dilemmas

If you are found to have a thyroid problem in later life, decisions about treatment are far from clear. On the one hand, as we've already observed, mild or subclinical hypothyroidism could boost your risk of developing heart disease whereas mild hyperthyroidism could increase the risk of heart failure. Screening for thyroid disease in the over-60s would, at the very least, enable those at risk to be more closely monitored. The incidence of subclinical hypothyroidism is higher among women in their 40s and 50s than in those in their 70s and 80s, suggesting that some women who had it at an earlier age did not live to reach old age perhaps as a result of their thyroid problems.

On the other hand, there are concerns that treatment may actually accelerate the development of thyroid problems. Doctors are particularly worried about the use of T3, which may overstimulate the heart in those with existing heart problems. It's a sad fact that women, especially elderly women, have been underrepresented in drug trials. The result is that far too little is known about how best to treat them, which has a knock-on effect for women with failing health in later life. It is vital that future researchers include older women in their study populations to eventually improve such women's health during what ought to be some of the best years of their lives.

# CHAPTER ELEVEN

# Questions and Conundrums

As you've read through this book, you have learned how attitudes towards thyroid problems are changing, and about some of the new management approaches being tried and some of the things you can do to help yourself. So where do we go from here? Will women with thyroid problems find it easier to be diagnosed in future? Are there more effective treatments on the horizon? Will it ever be possible to prevent thyroid problems occurring? This chapter examines some of the current controversies and debates.

## The Great Screening Debate

Given how common thyroid problems are, an important question remains: should doctors be more active in trying to track it down? In particular, should they be looking for and treating 'silent,' mild or subclinical thyroid disease? Is there, in fact, a case for all of us to have regular screening to look for signs of thyroid disease in much the same way as we now go for cervical smears or mammograms? And if so, at what age should screening begin – 19, 35, 40, 50? Should an abnormal blood test be considered a sign of early or 'pre-thyroid disease' and be treated, or is it of little or no significance? The debates have raged for two decades and show no signs of being resolved.

One argument for regular screening is that 'pre-thyroid' abnormalities such as raised or lowered levels of thyroid-stimulating hormone (TSH) or antithyroid antibodies can be clues that overt thyroid problems will develop later. It's estimated, for instance, that mild thyroid failure may develop into full-blown hypothyroidism at a rate of 5–26 per cent a year. Being aware of your risk could help you receive effective treatment for problems such as depression and overweight, which are often dismissed as insignificant or due to failure of willpower, sooner rather than later. It might enable women with unexplained infertility to be identified and treated earlier. And it could encourage women to take simple measures to strengthen the immune system by paying attention to diet and lifestyle or visiting a herbal practitioner, which could perhaps prevent overt thyroid problems from developing. Above all, routine screening could save countless women, like those whose stories appear in this book, from the misery of months and years of tiredness, anxiety, depression and vague feelings of unwellness. As one such woman, June, observes:

> *I feel I have lost what should have been the best years of my life to this illness. The whole of my thirties and early forties have been spent battling to feel well again, fighting constant fatigue, trying to lead something approaching a normal life.*

## Knock-On Effects

One of the most powerful arguments for screening is that both a mildly underactive thyroid and a mildly overactive one can raise the risk of more serious, possibly life-threatening, health problems such as heart disease and heart failure. These problems are particularly important for women in later life and tend to be less well treated in women than in men. Even minor degrees of thyroid underactivity can raise cholesterol levels. In fact, research in older women suggests that subclinical hypothyroidism may be a risk factor in its own right for heart attack of a similar magnitude to established risks such as high blood pressure, raised cholesterol, smoking and diabetes. Mild thyroid overactivity is associated with atrial fibrillation, an abnormal heart rhythm that is, in turn, associated with an increased risk of heart attack and stroke.

A further argument has to do with the effects of a silently underactive thyroid on the mind. Anxiety and depression are both linked to subclinical hypothyroidism, and at least one study has shown that mild hypothyroidism is also associated with panic attacks. It is also suggested that memory problems may be linked to an underactive thyroid. There is evidence to suggest that an overactive thyroid may boost the risk of developing dementia and Alzheimer's disease in persons over 55, especially if antithyroid antibodies are present. One study looking at the effects of giving thyroxine ($T_4$) to a small group of elderly patients with memory problems showed an improvement in mental function equivalent to a difference of 8.7 points in an IQ test.

Doctors are still in disagreement over the issue of osteoporosis, which strikes one in four women in later life. It has been found that women with silent hyperthyroidism due to multinodular goitre have significantly lower bone density. If left untreated, they can lose as much as 2 per cent of their bone mass a year.

However, not everyone with abnormal thyroid function test results or antithyroid antibodies goes on to develop full-blown thyroid disease. Moreover, the test results may be skewed because of the presence of other illness or certain medications and, in some cases, levels of TSH return to normal on their own. Is screening worth the inevitable worry that, at some unspecified point in the future, you might develop thyroid problems? And what action should be taken if the disease is not causing any problems? Should it still be actively treated and, if so, what is the basis for this decision? Or should the doctor just wait and see, as happens in the case of some mildly abnormal cervical smears?

## What Do Doctors Think?

With regard to mild, subclinical hypothyroidism particularly in the US, the current thinking seems to be moving towards the idea of screening every five years for women from age 35 onwards. This idea came on the heels of a study showing that periodic testing is as cost-effective as screening for other 'silent' conditions such as high cholesterol or high blood pressure. Given the potentially serious consequences of an underactive thyroid in pregnancy, a growing number of doctors in the USA and the UK believe that a thyroid test should become routine part of the standard battery of tests carried out antenatally.

Nevertheless, in general, UK doctors are more cautious. Guidelines produced in 1996 stated that – with a few exceptions – 'general testing of the population ... is unjustified ... even [in] high-risk groups such as women over 60 and those with a strong family history of thyroid disease'.

However, a number of experts believe that screening may be worthwhile in the following circumstances:

- a previous bout of postpartum thyroiditis (PPT)
- type 1 (insulin-dependent) diabetes (in pregnancy)
- unexplained infertility
- over age 40 with symptoms suggestive of an underactive thyroid
- depression that is not responding to treatment or manic-depression, with rapid swings from mania to depression
- certain chromosomal disorders such as Turner's and Down's syndromes
- raised cholesterol levels
- a personal or family history of thyroid disease
- a personal or family history of other autoimmune diseases
- autoimmune Addison's disease (adrenal failure caused by damage to the adrenal glands by autoantibodies)

## Treatment Trials

In the US, doctors recommend that treatment trials should be attempted for those with symptoms suggestive of an under-active thyroid, raised cholesterol levels or a goitre as well as in pregnant women and those with ovulatory problems.

In the UK, the consensus appears to be that it's worth try-ing to nip future problems in the bud if there's a high risk of developing overt hypothyroidism – in other words, treat individuals whose levels of thyroid-stimulating hormone are consistently above the upper limit, especially if antithyroid antibodies are present, as well as those who have been previ-ously treated for an overactive thyroid or goitre. Some doctors recommend treatment in the absence of symptoms for preg-nant women or those with ovulation problems and infertility.

But where does that leave those with TSH levels within the upper limit of normal, given the belief that the true upper limit for TSH may actually be lower than the standard reference range for normal? And what about those without antithyroid antibodies? More to the point, where does this advice leave the thousands of women who complain of symptoms but whose test results are 'normal'? Some doctors believe that, in these circumstances, a trial of thyroxine ($T_4$) for, say, three months may be warranted. However, many still prefer to wait and see. If your doctor takes this latter approach, then this should be discussed with you to give you the opportunity to explore the implications of treatment versus watchful waiting. You should also be offered an annual thyroid function test and a chance to discuss any symptoms.

The truth is that no one really knows what the long-term risks of taking $T_4$ are in cases of silent disease. There are concerns that it could aggravate undetected heart problems or thin the bones, leading to an increased risk of osteoporosis. There are also fears that it could trigger adrenal collapse in those with low levels of the stress hormone cortisol, or perma-nently disrupt the delicate feedback loop that operates between the hypothalamus, pituitary gland and thyroid. What's more, it could result in many perfectly well women taking possibly unnecessary drugs for the rest of their lives.

The picture for mild, silent hyperthyroidism is similarly cloudy. The American Association of Clinical Endocrinologists, for example, has concluded that subclinical hyperthyroidism associated with a goitre will usually need treatment. But many doctors disagree with the idea that those with subclinical hyperthyroidism with nodular disease or atrial fibrillation also warrant treatment, given the high risk of progression to overt hyperthyroidism in the former, and the risks of heart failure or stroke in the latter. However, there the consensus ends.

Some doctors think that, for people with symptoms such as chronic fatigue that could be due to an overactive thyroid, treatment with an antithyroid drug or radioiodine therapy should be tried. However, as with hypothyroidism, there are still many questions concerning the natural course of the disease – whether mild, subclinical hyperthyroidism is actually early disease, what are the chances of it causing more serious problems such as osteoporosis and heart failure, and what are the consequences of unnecessary treatment?

If you are young and have no heart problems, the doctor may suggest watching and waiting with repeat testing after three to six months. However, if there are symptoms such as nodules or lumps, a scan and a radioactive iodine test may be suggested as well as possibly a bone scan. And here again, the doctor should give you the opportunity to be fully involved in any decisions made.

Clearly, both underactive and overactive thyroid disorders need more research. In the meantime, it is up to you to be alert to the symptoms of thyroid disease and of other risk factors, such as your family history. You should see the doctor if you think you would benefit from testing and treatment. As Danish doctor Peter Laurberg put it in an article in the e-journal *Hot Thyroidology* on whether subclinical thyroid disease should be treated, 'In our opinion, a decision on watchful waiting should be taken by the informed patient, not by the doctor or the healthcare system'.

# Finding the Causes

The evidence that certain thyroid problems, such as Graves' disease, Hashimoto's thyroiditis, autoimmune thyroiditis and thyroid eye disease, are linked to a number of other autoimmune disorders has grown in the past few years. It has been found, for example, that persons with autoimmune diseases have small variations in their DNA – called polymorphisms in the medical jargon – that make them susceptible to Graves' disease as well as type 1 (insulin-dependent) diabetes and Addison's disease, which causes destruction of the adrenal glands.

Other researchers have been looking at one of the most puzzling questions in immunology: how does the body limit inflammation?

Research is also currently ongoing into the role that thyroid hormone – the bioactive $T_3$ form in particular – might play in diseases such as depression, chronic fatigue syndrome (CFS, or ME – myalgic encephalomyelitis in medical-speak), heart problems (including atherosclerosis, or furred, narrowed arteries, and heart failure) and dementia. The findings are leading to new thinking regarding the possible use of thyroid replacement therapy as a treatment for these diseases.

The relationship between genes and the environment continues to be a hot topic and might eventually lead to new treatments, although genetic glitches alone aren't enough to account for a faulty thyroid. For this, we must turn to the still only partly understood interplay between your genes, your environment and your lifestyle. As Professor Anthony Weetman, a UK endocrinology expert, said seven years ago, 'We're not going to find a single gene: it's more like a fruit machine – you'll get several lemons, both genetic and environmental'. This is indeed proving to be the case.

Over the past few years, research into genetics and molecular biology has provided fascinating insights into the origins of thyroid disease. A growing body of studies has identified the presence of thyroid antibodies as a risk factor for thyroid problems, with the suggestion that these are passed down in families through the genes, conferring a genetic susceptibility to thyroid disease especially in women. Researchers have now identified several 'candidate' genes for Graves' disease and a number of other thyroid disorders.

Strong evidence has also emerged that the genes involved in regulating T cells (white blood cells that patrol the body and repel any invaders) and cytokines (inflammation-causing chemicals) are important factors in dictating disease development in those who are susceptible. A question mark remains over whether an aberration on chromosome 21, the chromosome associated with both Alzheimer's disease and Down's syndrome, could influence the development of autoimmune thyroid disease.

The essential function of genes is to regulate the behaviour of our cells. That's why they are often referred to as the 'blueprints of life'. Research into the mechanisms by which thyroid hormone ($T_3$ and $T_4$) might regulate the genes involved in instructing organs and tissues – like the brain and central nervous system, heart and liver – how to behave could open up the possibility of gene therapy to fix the faulty genes, thereby treating the root cause of thyroid problems rather than only its symptoms.

## What Part Might Environment Play?

As we've seen throughout this book, genes may load the gun of autoimmunity, but factors in the environment are required to pull the trigger. One such triggering factor may be infections. Although no specific culprits have yet been identified, it is thought that infectious agents may trigger autoimmunity by disguising themselves as 'self' to evade the immune system or by activating innocent bystanders in the immune army.

Unfortunately, attractive though the suggestion is, there is still no hard evidence that infections are involved in autoimmune thyroid problems, but it's certainly an area that warrants further research. Given that many aspects of modern life – such as poor digestion and absorption, poor diet, toxins, drug side-effects and stress – can all lower immune resistance and make us prone to minor infections, it could be that bolstering the immune system by eating a healthy diet, and getting sufficient exercise and relaxation, will make us more resistant to infection and help prevent the development of at least some thyroid problems.

The links between stress and thyroid problems, especially Graves' disease, have been strengthened since the first edition of this book. A number of experts surmise that prolonged or persistent stress may somehow affect the circuit of messenger chemicals between the hypothalamus, pituitary and adrenals – the glands responsible for our body's stress responses – leading to permanent alterations in immunity. Stress is also implicated in difficulties conceiving, perhaps again because it somehow disrupts the hypothalamus. It could be Nature's way of ensuring that women don't become pregnant under conditions that are environmentally hostile to both mother and baby. Whatever the truth of it, it seems likely that taking steps to manage stress may also be good for your thyroid.

## Environmental Poisons

Another question mark lies over whether environmental toxins, such as cigarette smoke, pesticides, food additives and other industrial chemicals, disrupt the body's delicate hormonal balance and contribute to the development of autoimmune thyroid and other problems. Clinical ecologists, doctors who are concerned with the effects of the environment on disease, have long claimed that such factors are to blame for a vast number of illnesses, although conventional doctors have tended to dismiss such claims.

Nevertheless, there is now more recent evidence supporting an association between smoking and Graves' and thyroid eye disease, and even non-autoimmune hyperthyroidism. There are several theories as to how smoking may contribute to thyroid problems: cigarette smoke may directly irritate the thyroid; nicotine may indirectly stimulate the nervous system;

smoking may alter the structure of the thyroid-stimulating hormone (TSH) receptor, making it more receptive to auto-antibodies; or smoking may somehow change the responses of the immune agents known as T cells. Whatever the truth of the matter, these are all good reasons for doing everything you can to give up smoking if and when you feel able to.

And what about other environmental pollutants, such as pesticides and other industrial chemicals? As far back as 1991, Professor Weetman reported that certain products derived from coal tar (anthracene) sparked off autoimmune thyroid problems in certain genetically susceptible strains of rats. Coal-tar derivatives are used in dyes and food additives, so could the environmentalists be right after all? No one knows, but Weetman commented:

> *It seems likely that a combination of genetic,*
> *constitutional and environmental factors initiate*
> *autoimmune thyroiditis [although] the relative*
> *contribution of each of these, and almost certainly*
> *of undiscovered factors, is unknown.*

The good news is that something is being done to see what, if any, effect pollution may have on the hormonal system. In 2002, the European Union announced the intention to pour 20 million euros into research designed to clarify the effects of potential 'endocrine disruptors', although it is likely to be some time before any definitive answers are found. These chemicals are thought to exert their harmful effects by mimicking or interfering with the body's natural hormones. To some extent there's not a lot an individual can do to avoid such pollutants: it has to be the responsibility of governments to cap toxic emissions and legislate on the chemicals used in industry. However, it makes sense to avoid pollution as far as possible by thinking about how and where you live, and considering organic foods.

Such research could greatly improve the quality of life for those who have autoimmune thyroid disease, one of the most important causes of thyroid problems. In particular, it could also settle any disputes over the origins of and best treatment for those with thyroid eye disease, a condition that remains an enigma.

## Future Positive?

So, what of the future? And what do the various debates and new research mean to women with thyroid problems? The experts clearly still have much more to learn, although there are some promising developments.

On the treatment front, there seems to be a small but distinct swing towards the use of a cocktail of $T_3$ and $T_4$, rather than $T_4$ alone, which could bring striking improvements for some women with thyroid disorders, especially those with depression, lethargy and loss of mental sharpness. There may also be improved surgical procedures, such as the use of 'keyhole' surgery – done through an endoscopy tube and requiring only a small skin incision – for the removal of thyroid nodules. The growing interest in autoimmune diseases should also have a beneficial effect on doctors' awareness and understanding of thyroid problems.

Genetic and environmental research will almost certainly lead to more sophisticated treatments designed to stem the disease process itself, especially for those with an overactive thyroid, rather than the somewhat crude treatment of symptoms used at present.

Treatments might include developing monoclonal antibodies – artificial antibodies designed to home in on particular foreign cells – or 'suicide genes' designed to kill tumour cells in thyroid cancer. Other treatments could involve antagonists – drugs that work by preventing antigens from binding to receptors and so block the cell's response.

A further possibility is, at some time in the future, the development of a vaccine to prevent thyroid problems in those at risk. As Professor Weetman explains:

*If you take the T cells that cause disease and treat
them in such a way that it prevents them causing
disease, for example, by irradiating them and then
reintroducing them into the body, that 'vaccine'
will prevent the disease occurring. Alternatively,
receptors that recognize the T cells could be
produced as a 'vaccine'. Even more subtly, if you
take bits of protein from the thyroid, you can trick
the immune cells into not responding.*

Any such vaccine would need to be given right at the
beginning of the disease, something that could be difficult to
pinpoint in view of the insidious way in which thyroid disease
tends to creep up on a person. The only way this could be
achieved is if it were possible to pinpoint those especially at
risk because of a family history of thyroid problems – which,
of course, brings us back to the screening debate.

Such treatments are still, of course, some way in the future,
but the fact that thyroid disease is being taken more seriously
should mean that potentially more effective treatments are on
the horizon. In the meantime, if you've got this far, you will
realize that thyroid disease doesn't have to rule your life –
there are plenty of ways in which you can help yourself.
Experimenting with some of these will enable you to feel a
renewed sense of energy and vigour so that you can lead a
healthy, active existence and regain your zest for life.

# Glossary

**Antibody**
A protein manufactured by certain types of white blood cells (lymphocytes) that destroy foreign proteins. In autoimmune thyroid disorders such as Hashimoto's and Graves' disease, antibodies form against the body's own cells.

**Antigen**
A substance that is foreign or different from the body's own proteins that triggers an immune response. Sometimes, harmless substances or proteins that are not foreign are mistaken as a potentially harmful agent, causing the immune system to overreact, as in the case of an allergy, or to turn against itself, as in autoimmunity.

**Autoantibody**
An antibody directed not against an invader, but against the self.

**Autoimmunity**
The process by which the body turns against itself and produces antibodies to destroy its own tissues.

**B cells**
White blood cells that help protect the body against infection.

## Graves' disease
Autoimmune thyroid disease caused by autoantibodies directed against thyroid-stimulating hormone (TSH) receptors leading to overactivity of the thyroid.

## Graves' ophthalmology
Thyroid eye disease, also known as 'thyroid ophthalmopathy', 'dysthyroid orbitopathy', thyroid-associated orbitopathy (TAO); also an autoimmune disease.

## Hashimoto's thyroiditis
Autoimmune thyroid disease thought to be caused by antibodies against thyroid peroxidase (TPO) and thyroglobulin (TG) leading to an underactive thyroid.

## Hyperthyroidism
Biochemical state of having an overactive thyroid.

## Hypothalamic–pituitary–thyroid axis
Connections that link the three major organs involved in regulating thyroid function.

## Hypothalamus
Organ of the brain that regulates pituitary function by producing thyrotropin-releasing hormone (TRH).

## Hypothyroidism
Biochemical state of having an underactive thyroid.

## Immunoglobulin
Another name for an antibody; a type of protein, found in the blood and tissue fluids, produced by cells of the immune system called B lymphocytes. They bind to foreign antigens and, in the case of bacteria, viruses and other microorganisms, this binding is a crucial event in destruction of the microorganism. In the case of autoimmune thyroid disease, it is a key event in the destruction of the thyroid.

**Monodeiodination**
Removal of an iodine atom from thyroid hormone that converts $T_4$ (thyroxine) into $T_3$ (triiodothyronine).

**Myxoedema**
Clinical condition, including dry, waxy swelling of the skin, and facial changes such as a thickened nose and swollen lips, caused by severe hypothyroidism.

**Negative feedback loop**
Control mechanism involved in triggering or inhibiting the production of hormones.

**Pituitary**
Master gland, found at the base of the brain, that regulates the function of the thyroid and adrenal glands, and ovaries and testicles.

**Postpartum**
After birth.

**Receptor**
A protein structure found on the surface of a cell onto which hormones and other messenger chemicals attach in order to exert their action inside the cell. Autoantibodies can also lock onto receptors, causing the body to turn against itself.

**Thyroid ablation**
Removal of the thyroid by surgery or its destruction by radioiodine therapy

**Thyroid-binding globulin (TBG)**
The main transport protein for thyroid hormone in the bloodstream.

**Thyroidectomy**
Surgical removal of the thyroid.

**Thyroid hormone (TH)**
Combination of $T_3$ and $T_4$ produced by the thyroid.

**Thyroiditis**
Inflammation of the thyroid.

**Thyroid nodule**
A knot or lump of tissue found in the thyroid.

**Thyroid peroxidase (TPO)**
An enzyme that acts as a catalyst for the production of thyroid hormone within the thyroglobulin molecule.

**Thyroid-stimulating hormone (TSH)**
A protein hormone produced by the pituitary gland that stimulates the release of thyroid hormone by the thyroid.

**Thyrotoxicosis**
The physical and mental effects of an overactive thyroid. The state of being hyperthyroid.

**Thyrotropin-releasing hormone (TRH)**
A messenger chemical produced by the hypothalamus that triggers the secretion of thyroid-stimulating hormone (TSH) by the pituitary.

**Thyroxine**
$T_4$, one of the two hormones comprising thyroid hormone.

**Triiodothyronine**
$T_3$, the bioactive hormone comprising thyroid hormone with thyroxine ($T_4$).

### TSH receptor
Protein on the surface of thyroid cells onto which thyroid-stimulating hormone (TSH) binds to stimulate the production of thyroid hormone.

### Tyrosine
An amino acid (building block of protein) found in thyroglobulin. When iodine combines with tyrosine through an enzymatic reaction, thyroid hormone is produced.

# Resources

## Thyroid Organizations

CANADA
Thyroid Federation International
96 Mack Street
Kingston
Ontario ON K7L
E-mail: tfi@on.aibn.com
Website: www.thyroid-fed.org
Useful information on thyroid problems and links to other organizations worldwide. Good source of information on how to start a thyroid patients' group

Thyroid Foundation of Canada
P.O. Box/CP 1919 Stn Main
Kingston
Ontario K7L 5J7
Website: www.thyroid.ca

UK
British Thyroid Foundation (BTF)
P.O. Box 97
Clifford Wetherby
West Yorkshire LS23 6XD
E-mail: info@btf-thyroid.fsnet.co.uk
Website: www.british-thyroid-association.org

Affiliated to the doctors' British Thyroid Association. Produces a useful newsletter and simple leaflets covering a wide range of thyroid problems

Thyroid UK
32 Darcy Road
Clacton-on-Sea
Essex CO16 8QF
Website: www.thyroiduk.org
Produces a thought-provoking newsletter and information pack. Useful source of information about animal thyroid preparations, and the potential links between thyroid problems, myalgic encephalomyelitis (ME, or chronic fatigue syndrome, CFS) and adrenal failure. Maintains a list of doctors interested in an alternative approach to thyroid problems

Thyroid Eye Disease Association
Solstice
Sea Road, Winchelsea Beach
East Sussex TN36 4LH
E-mail: tedassn@eclipse.co.uk
Useful newsletter and leaflets specifically for people with thyroid eye disease. Associated with British Thyroid Association and founder member of Thyroid Federation International

Changing Faces
1 & 2 Junction Mews
London W12 1PN
E-mail: info@changingfaces.co.uk
Charity dedicating to helping people manage their altered appearance more successfully. Practical advice and support for rebuilding self-confidence

Institute for Complementary Medicine (ICM)
P.O. Box 194
London SE16 7QZ
E-mail: icm@icmedicine.co.uk
Website: www.icmedicine.co.uk

British Complementary Medicine Association
P.O. Box 5122
Bournemouth BH8 0WG
E-mail: web@bcma.co.uk
Website: www.bcma.co.uk

USA
National Graves' Disease Foundation
P.O. Box 1969
Brevard, NC 28712-1969
E-mail: ngdf@citcom.net
Website: www.ngdf.org

Thyroid Foundation of America
410 Stuart Street
Boston, MA 02116
E-mail: info@tsh.org
Website: www.allthyroid.org

ThyCa: Thyroid Cancer Survivors' Association
P.O. Box 1545
New York, NY 10159-1545
E-mail: thyca@thyca.org
Website: www.thyca.org

Light of Life Foundation (for patients with thyroid cancer)
32 Marc Drive
Englishtown, NJ 07726
E-mail: info@lightoflifefoundation.org
Website: www.lightoflifefoundation.org

USEFUL WEBSITES
Website: www.MyThyroid.com
An evidence-based patient-centred website set up by Canadian endocrinologist Dr Daniel J. Drucker. Useful source of medical information to enable you to keep up with the latest research and follow the debates on thyroid problems

Website: http://thyroid.about.com
This website of author Mary Shomon is a collection of daily updates including useful information and news on all aspects of thyroid problems, the latest debates and complementary therapies

## Further Reading

THYROID PROBLEMS
*Living Well with Hypothyroidism. What Your Doctor Doesn't Tell You ... That You Need to Know* by Mary J. Shomon (Quill, 2001)
*Thyroid Power: 10 Steps to Total Health* by Richard L. Shames, MD, and Karilee Halo Shames, RN, PhD (HarperCollins, 2001)
*The Thyroid Solution. A Revolutionary Mind–Body Program That Will Help You* by Ridha Arem, MD (Ballantine Books, 1999)
*The Thyroid Sourcebook for Women* by M. Sara Rosenthal (Lowell House, 1999)
*Your Thyroid: A Home Reference* by Lawrence C. Wood, MD, David S. Cooper, MD and E. Chester Ridgway, MD (Ballantine Books, 1995)

COMPLEMENTARY THERAPIES, DIET AND EXERCISE
*The Complete Guide to Integrated Medicine: Complementary Therapies Combined With Medical Science* by Dr David Peters and Anne Woodham (Dorling Kindersley, 2000)

*Fitness For Life Manual* by Matt Roberts (Dorling
    Kindersley, 2002)
*The Natural Health Handbook For Women: The Complete
    Guide to Women's Health Problems and How to Treat
    Them Naturally* by Marilyn Glenville, PhD (Piatkus,
    2001)
*Nutritional Medicine: The Drug-Free Guide to Better Family
    Health* by Dr Stephen Davies and Dr Alan Stewart
    (Trans-Atlantic Publications, 1987)

# Index

# Make
# www.thorsonselement.com
## your online sanctuary

Get online information, inspiration and
guidance to help you on the path to physical
and spiritual well-being. Drawing on the integrity
and vision of our authors and titles, and with
health advice, articles, astrology, tarot, a
meditation zone, author interviews and events
listings, www.thorsonselement.com is a great
alternative to help create space and peace
in our lives.

So if you've always wondered about practising
yoga, following an allergy-free diet, using the
tarot or getting a life coach, we can point you
in the right direction.

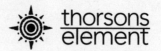

www.thorsonselement.com